ENVIRONMENTAL REGULATION OF INDUSTRIAL PLANT SITING

**The Conservation Foundation
Board of Trustees**

Ernest Brooks, Jr.
  Chairman
William H. Whyte
  Vice Chairman
T.F. Bradshaw
John A. Bross
Louise B. Cullman
Gaylord Donnelley
Maitland A. Edey
David M. Gates
Phillip G. Hammer
Walter E. Hoadley
William F. Kieschnick
William T. Lake
Richard D. Lamm
Melvin B. Lane
David Hunter McAlpin
Ruth H. Neff
Eugene P. Odum
Richard B. Ogilvie
Walter Orr Roberts
James W. Rouse
Anne P. Sidamon-Eristoff
George H. Taber
Henry W. Taft
Pete Wilson
Rosemary M. Young

William K. Reilly
  President

The Conservation Foundation is a nonprofit research and communications organization dedicated to encouraging human conduct to sustain and enrich life of earth. Since its founding in 1948, it has attempted to provide intellectual leadership in the cause of wise management of the earth's resources.

ENVIRONMENTAL REGULATION OF INDUSTRIAL PLANT SITING
How To Make It Work Better

Christopher J. Duerksen

The Conservation Foundation
Washington, D.C.

**Environmental Regulation of Industrial Plant Siting:
How To Make It Work Better**
© 1983 by The Conservation Foundation. All rights reserved. No part of this book may be reproduced in any form without the permission of The Conservation Foundation.

Cover design by Sally A. Janin
Typography by Rings-Leighton, Ltd., Washington, D.C.
Printed by R.R. Donnelley & Sons Company, Harrisonburg, Virginia

The Conservation Foundation
1717 Massachusetts Avenue, N.W.
Washington, D.C. 20036

Library of Congress Cataloguing in Publication Data
Duerksen, Christopher J., 1948–
   Environmental regulation of industrial plant siting. Includes bibliographical references and index.
   1. United States—Industries—Location—Environmental aspects.
   2. Environmental policy—United States.
I. Title.
HC110.D5D83   1983   338.6'042'0973   83-20901
ISBN 0-89164-078-9

Contents

FOREWORD by *William K. Reilly*	ix
PREFACE	xiii
EXECUTIVE SUMMARY	xix

PART I: THE NEW RULES OF THE GAME

1 / SITING INDUSTRIAL FACILITIES: THE RULES CHANGE	1
The Siting Process Yesterday: Paradise Lost?	1
Theory versus Practice	2
Siting in the 1960s	2
Kelly Springfield	2
BASF	4
The Siting Process Today: Out of the Garden of Eden	7
Industry Perspective	8
Determination of Need/Appropriation of Capital	8
Search for a Plant Site	10
Plant Design, Construction, and Operation	11
Regulatory Perspective	11
Preapplication Permit Identification	12
Preparation of Permit Applications	13
Submission and Processing of Applications	13
Conclusion	15
References	15
2 / ENVIRONMENTAL LAWS: ROADBLOCKS TO ECONOMIC GROWTH?	17
Celebrated Siting Disputes	18
Dow versus California	18
The Project	18
The Economic, Social, and Political Context	19
The Environmental Regulatory Process	20
The Outcome	21
Analysis of Regulatory Difficulties	22
SOHIO's Pactex Project	26
The Project	26
The Environmental Regulatory Process	27
Environmental Concerns	28
The Outcome	29

Analysis of Regulatory Difficulties	31
Significance of Regulatory Delays	33
Oil and Oysters—The HRECO Oil Refinery	35
The Project	35
The Environmental Regulatory Process	36
The Outcome	38
Analysis of Regulatory Difficulties	39
New Rules of the Game	43
References	44

PART II: PROBING THE HEADLINE CONCERNS: MYTHS AND REALITIES

3 / THE COST-COMPETITION FACTOR	**49**
Myth: Environmental Quality Regulations Cause Industries to Flee to Other Countries	**50**
Exploring the Issues	52
Summary	56
Myth: Environmental Lures Lead to Interstate Industrial Flight	**56**
Exploring the Issues	58
Summary	71
References	71
4 / THE PERMITTING-MAZE SYNDROME	**79**
Myth: Red Tape is Strangling Industrial Development	**81**
Exploring the Issues	82
Summary	88
Myth: Other Countries Are Better at Reconciling Environmental Regulation with Industrial Development	**88**
Exploring the Issues: Foreign Case Studies	89
Barlöcher Chemical	89
STEAG	95
European and U.S. Siting Parallels	100
Summary	102
No Need for Reform?	103
References	103

PART III: RESPONSES TO THE NEW REALITIES

5 / ATTEMPTS TO IMPROVE THE RULES: REFORMS OF THE 1970s	**109**
Streamlining the Process: One-Stop Permitting and Consolidation	109

Georgia's Consolidation Attempt	112
Truncating the Process: Ignoring Certain Laws, Governments, and Citizens	115
Florida	118
Washington	119
Can Truncation Work for Industrial Siting?	120
Site Selection by Government Agencies	123
Maryland	124
San Francisco	125
Europe	125
Hog-Tying the Regulators	126
Putting It All Together: A Strategy of Cooperation	130
How the Colorado Joint Review Process Was Conceived	130
AMAX Tests the CJRP	131
Rating the CJRP	135
Conclusion	141
References	142
6 / RESPONSES TO THE REAL PROBLEMS: PERMIT SYSTEM INNOVATIONS FOR THE 1980s	149
The Role of Government	150
Coordination through Better Management	150
Lead Agencies	151
Expert Project Teams	152
Joint or Consolidated Hearings	154
Project Managers and Escort Services	156
Interagency Mediating Bodies	157
The Need for Predictability	158
Project Decision Schedules	159
Project Tracking Systems	161
Grandfather Clauses	162
Regulatory Personnel Problems	165
The Challenge to Industry	168
Early Environmental Impact Assessment	169
Environmental Reconnaissance Statements	172
Improved Project Organization and Project Management	174
Early Public Involvement	176
Greater Attention to Indirect, Quality-of-Life Impacts	178
Adoption of New Ways to Avoid or Settle Disputes	180
Mediation	182
Mitigation	183

Closing the Communication Gap	184
Adequate Information for Regulators	185
Meshing the Regulatory Process with Project Planning	185
Trade Secrets and Land Speculation	187
Scoping	189
Helping Local Governments	190
More Information for Project Proponents	192
Identifying Off-Limit and Acceptable Sites	193
Compiling Permit Registers and Guides	196
References	197
7 / SOME CLOSING THOUGHTS	205
Tips for Regulatory Reformers	205
Maintain the System's Integrity	205
Provide Necessary Checks and Balances	206
Involve All Affected Parties	207
Provide Adequate Financial and Administrative Support	208
Maintain High Visibility	209
Take a Comprehensive Approach	210
Be Persistent	211
Continuing Challenges	211
References	214
APPENDIX A: RANKING OF STATES' ENVIRONMENTAL CONTROLS	218
APPENDIX B: THE REMODELED COLORADO JOINT REVIEW PROCESS	230
INDEX	233

Foreword

Industrial development in the United States benefits not only the American—and, indeed, the world—economy but also the environment. Replacing old industrial plants with new capacity offers the promise of reducing pollution associated with industrial output. Failure to do so may increase the technological obsolescence of U.S. industry, exacerbating an already serious economic situation and braking progress in cleaning up our country's environment.

During the past decade, the United States has witnessed impressive gains in cleaning up the nation's air and water. But these gains have been accompanied by a recurring complaint: the same regulations that have led to environmental progress have been accused of stifling needed investment in industry and eroding our competitive position internationally. The regulatory system, it has been alleged, does not work because it has been put together and implemented by people who have not understood its effect on industry, particularly on planning and building big manufacturing and energy projects.

A careful review of these complaints indicates that some of them are quite correct, although the extent of adverse impact, and the role of environmental regulations relative to other factors inhibiting U.S. economic competitiveness, have tended to be exaggerated. It is worth considering the implications of this acknowledgment for environmental policy. Most people, including environmentalists, do not fully understand how firms develop industrial projects and just what role environmental-quality laws play in that process. Particularly worrisome have been growing demands that environmental standards be lowered to facilitate needed industrial development, even in the absence of evidence that such standards have caused delays and impeded the siting of new industry.

To address these concerns, The Conservation Foundation has conducted over the past four years an Industrial Siting Project. Our goal has been to understand how the siting process works, and to identify ways to improve the environmental regulatory system without sacrificing environmental quality. Generous support for the project has come from The German Marshall Fund of the United States, The Ford Foundation, and the Richard King Mellon Foundation, to whom we express our gratitude. We are also grateful to the Foundation's donors of unrestricted support, for we supplemented our project gifts in order to add several important elements to the research.

Our project team brought together diverse backgrounds and skills. Christopher J. Duerksen, an attorney who served as principal investigator for the project, previously represented both state and local governments and industry while practicing land-use and environmental law in Chicago. Dr. Robert G. Healy, an economist, has done work in the field of industrial location and authored one of the leading books on state land-use regulation. Michael Mantell, also an attorney, has had extensive experience in environmental law in local government, and Dr. Richard A. Liroff contributed a political scientist's skills in analyzing complex organizational and institutional issues. H. Jeffrey Leonard, also a political scientist, brought to the project a unique blend of knowledge about international economic development and environmental affairs. In addition, because of our continuing belief in the value of learning from the experiences of other Western nations, and because of claims that those countries were building new plants more quickly than was the United States, we joined forces with the International Institute for Environment and Society (Berlin) to examine industrial siting and environmental regulation in Europe.

The project team carried out original statistical analysis of investment trends, reviewed existing studies, and conducted many interviews with industry and government officials. This work was augmented by a series of field studies of several major siting disputes. The team investigated instances where industry ran into difficulties—though we knew the cases were not typical—to identify real problems in the system, as distinguished from theoretical predictions or popular perceptions.

What we have learned is surprising and at same time reassuring. When we began the project, our eyes were cast, quite frankly, toward problems generally associated with government regulation—overlapping and contradictory permit reviews, changing laws and regulations, and never-ending judicial review. We have found that there are a number of ways government can improve, but we have concluded that government cannot do it alone. Companies, too, have an essential role to play. They have an obligation to understand better the demands on regulators and to improve the way they plan and execute big industrial projects. Without improvements by the private sector, true relief will never come.

Perhaps most important, the United States must avoid the lure of panaceas that promise to cure all of our regulatory ills quickly and with little pain. Experience shows clearly that the path to success lies in "quiet" reforms that do not ignore citizens, override or weaken laws, or preempt government agencies.

We are encouraged by the fact that companies and government agencies are adapting, learning from their experiences, and overcoming the teething pains of the 1970s, when traditional industrial expectations clashed with untried environmental policies. Already the most innovative government agencies and the most progressive corporations have begun exciting initiatives that hold promise to improve not only the way environmental laws work but also their effectiveness in protecting the environment.

Our Industrial Siting Project has produced several publications that we hope will help lead to a better understanding of the links between industrial growth and environmental regulation. In addition to this book, the project has produced several articles, as well as reports, including one on the environmental implications of industrial growth in the 1980s (by Robert G. Healy) and a forthcoming analysis of the international aspects of industrial location and environmental regulation (by H. Jeffrey Leonard). We also published a detailed case study of the celebrated Dow siting dispute in California by Christopher J. Duerksen. Also forthcoming is a series of foreign case studies by Mr. Leonard.

Neither economic recovery nor the long-term health of American industry need be hampered by efforts to ensure a clean and livable environment. A satisfying quality of life—to which environmental regulation can contribute so much—will be more important than ever in attracting the industries that will provide future jobs and economic growth.

William K. Reilly
President
The Conservation Foundation

Preface

The environmental decade of the 1970s came to a close with an increasing number of people asking hard questions about the economic impact of environmental laws and regulations. A chorus of critics, with visions of car queues at gas stations in their heads, claimed these laws were hobbling the nation's search for secure energy supplies. Other people were feeling some very un-American gnawings: the United States no longer seemed to be the number one industrial power. Many sectors of the economy that had been mainstays—steel, autos, mining, and smelting—appeared unable to compete with firms from powers like West Germany, Japan, and France. Leaders in these sectors pointed an accusing finger at the panoply of environmental regulations that domestic firms had to comply with. Celebrated battles over the siting of big energy and industrial[1] facilities—the trans-Alaskan pipeline, oil refineries on the East Coast, and a Dow Chemical complex in California—seemed to confirm the worst fears: the cost and complexity of all these new environmental-quality laws and regulations were thwarting needed projects here and driving investment to friendlier climes.

This set the stage for the Foundation's Industrial Siting Project, of which this book is a product. I began work in early 1979 with help from Robert Healy, an economist, and H. Jeffrey Leonard, a political scientist. Our sights were fixed on the "headline" concerns that were attracting much media attention, such as whether U.S. firms might flee to other countries to avoid stringent environmental laws here. A 1978 study alleged just that, warning of a wholesale movement of industry abroad. We also began studying charges that the permitting system was the modern-day equivalent of a Gordian knot with no clues to the unraveling. Was regulatory red tape really killing projects? We were fortunate to team up with a highly respected West German organization, the International Institute for Environment and Society (IIES), which, under the direction of Meinolf Dierkes and Gabriel Knödgen, began researching the same issues in Europe.

After several months of reviewing existing literature, analyzing industry investment patterns, and interviewing industry officials, regulators, and environmentalists, we became convinced that, although complying with environmental and land-use regulations was costly, it was not nearly so onerous as was often portrayed. Yet we knew that regulations were causing industry some headaches in the siting of new facilities, and that industry's headaches didn't necessarily translate into environmental protection. Moreover, we found a startling lack of comprehension among environmentalists and regulators about how big industrial facilities are planned and built. By the same token, we discovered that industry project planners did not always understand how the regulatory process worked and what peculiar demands faced those charged with implementing the new laws.

The logical question was "What next?" The overwhelming recommendation from 75 experts from industry, government, academia, and public-interest groups who gathered at a Conservation Foundation conference on industrial siting in June 1979 was to augment and go beyond our data analyses and interviews and to examine how the siting regulatory process worked—or didn't work—in practice, both in the past and today.

A number of participants wanted more information about how industry sited projects and how governments regulated them. Many others felt that case studies of location decisions, selectively chosen, would be particularly useful. As one, Stanley Dempsey (a vice-president of AMAX, Inc.), later wrote: "I've followed the research in business schools over the years and have been surprised that there have not been more case studies involving the management of environmental conflict, because certainly there has been enough of it in the last 10 years or so. I could teach an entire MBA course out of one big environmental conflict."[2] We agreed with Dempsey's advice, convinced that unless both industry officials and environmentalists had a better understanding of the impact of environmental regulations on the plant-siting process, the chances were slim that useful recommendations for streamlining the process could be made and that red tape could be cut to head off more virulent strains of reform.

As our next step, we solicited suggestions for case studies in the United States and abroad that could highlight problems connected with environmental and land-use regulations. We looked for controversial cases on the assumption that any existing problems with the siting process would surface in them. After reviewing hundreds of possibilities, we winnowed the list down to nine cases that seemed to illustrate the problems at their worst—three cases in the United States, three in West Germany, and three in France. All were celebrated and highly publicized disputes, often cited by industry as the worst of the lot.

I then worked with CF staff members Richard A. Liroff, a political scientist, and Michael Mantell, an attorney, to study the three U.S. cases in great detail. Participants on all sides of the disputes were interviewed, sites were visited, and secondary literature about each was reviewed. In Europe, IIES received assistance from Michael Pollak, a political scientist, and from Alain Davezac, of the Fondation pour le Cadre de Vie based in Paris. At the same time, H. Jeffrey Leonard began field work in Ireland, Spain, and Romania in an attempt to document or disprove charges that these and other rapidly industrializing countries were luring industry with the promise of weak environmental regulation, and Robert Healy began examining the environmental implications of industrial growth during the 1980s.

This field work was supplemented in several ways. First, we surveyed existing facility-siting case studies, including accounts both of successes and of projects that had been stopped or delayed, and then talked with the authors of many of these studies to determine what lessons could be learned from them. All totaled, we reviewed over 100 siting cases. Second, we interviewed numerous people across the country in government, industry, academia, and public-interest groups to get their perspectives on the project planning and siting process. Finally, we studied several cases in which innovative approaches to siting new industrial plants appeared to offer hope of a more efficient and effective system.

This book is a product of our work. It summarizes research performed by Leonard, Healy, and Knödgen, as well as the U.S. and foreign case studies, and presents our conclusions on the best path to remodeling the siting regulatory process.

The national scene has changed dramatically since we began our work. In 1978, the economy was rebounding strongly from earlier oil price shocks. Industrial growth prospects were bright, energy exploration was booming, and high-tech manufacturing firms were multiplying like rabbits. However, all this growth was accompanied by bottlenecks in the permitting system and an increasing number of siting disputes. In addition, there were worries that environmental regulations might stifle economic growth. In Wisconsin, for example, aluminum foundries warned that a convoluted, time-consuming air pollution permit system might force them to locate in Canada so that they could meet growing demands by carmakers in a timely fashion.

Energy production was a particular concern in 1979, and in July of that year President Carter proposed an energy mobilization board, with powers to override environmental laws and to expedite so-called priority energy projects. Environmentalists teamed with state and local officials to defeat that initiative.

Today, the industrial-development picture is far different. Regulatory reform is still of great interest, but in a markedly altered context. Environmental-quality regulations have been accused of hobbling recovery from a deep recession. Both Congress and the Reagan Administration have proposed their own brands of cure-all reforms, with many in and out of government firmly believing that nothing should be allowed to stand in the way of needed industrial revitalization.

This turn of events, we believe, makes the results of our work even more timely. The issue clearly is not *whether* industrial growth will occur, but *how*. Our work indicates that environmental-quality regulations need not stand in the way of this growth, nor need they be sacrificed on the altar of recovery.

The United States is emerging from its deep recession, and projections are that currently excess industrial capacity will not be fully utilized until 1985 at the earliest. In other words, few new plants will be built in the next year or so. This grace period before the environmental review process again clashes with industrial growth—be it construction of new plants or modernization of old ones—can be a do-nothing era or a time to begin carefully implementing reforms.

We believe that for all its ill effects the recent recession offers the nation a golden opportunity—breathing room for a new generation of reformers to proceed carefully in their revamping of the industrial-siting regulatory process. Our studies have persuaded us that a series of "quiet" reforms offers the best hope for effective, efficient solution. After examining specific examples of industrial-siting problems, as well as several myths surrounding environmental regulations, this book details those suggested quiet reforms.

Many people made important contributions to this project through case-study interviews, by sharing their research results and insights, through prodding, cajoling, and debate with project staff. Notable among these were Dean Misczynski, formerly of the California Governor's Office of Planning and Research and Stuart Sessions of the U.S. Environmental Protection Agency, who were particularly gracious in helping us analyze the regulatory process from the regulator's point of view. They have done some of the most thoughtful work on regulatory reform we came across. Arthur Biddle of AMAX, Inc., and Barbara Goldsmith of Environmental Research and Technology schooled us in the way companies plan and build industrial projects and reviewed the draft manuscript, as did Professors Howard Stafford of the University of Cincinnati and Thomas Gladwin of the New York University School of Business Administration. Both Stafford and Gladwin are well known for their extensive work in the field of industrial location theory and practice. Angelo Siracusa of the Bay Area Council lent valuable assistance in helping us present our research results at a May 1983 conference in San Francisco, and the continuing support of Marianne Ginsburg of The German Marshall Fund kept the project moving forward.

Perhaps the most gratifying aspect of the project was the cooperation and invaluable advice and insights I received from my colleagues here at The Conservation Foundation and at IIES in Berlin. My deepest thanks to H. Jeffrey Leonard, Robert G. Healy, Michael Mantell, Richard A. Liroff, and Gabriele Knödgen. Many others on our staff also contributed in a variety of ways. William Heinemann-Ethier and Angela Jordan both provided invaluable research assistance, as did librarian Barbara Rodes. Linda Roll and Laura O'Sullivan coordinated the administrative end of this project, leaving me free to think and write, while Tony Brown typed the many drafts of the book. At times, I felt Jack Noble, Robert McCoy, Bethany Brown, and Bradley Rymph (as tough a set of editors as one might care to meet) were ghostwriting for me. Thanks to them for their care and nurturing of the manuscript. If it is any consolation to my friends in industry, they should be able to get their plants built more quickly than I produced this book.

Christopher J. Duerksen
October 1983

REFERENCES

1. Throughout this book the terms *industrial* and *manufacturing facilities* are used interchangeably. Technically, the term *industry*, as used by the Departments of Labor and Commerce, includes many sectors ranging from ones with few, if any, major pollution or land-use problems (for example, wholesale and retail trade) to those that are normally associated with such environmental problems as air and water emissions (manufacturing sectors like mining and chemical production).

This study has focused on the latter, as well as on non-manufacturing sectors, such as non-nuclear electric utilities, that are often involved in siting disputes and have significant pollution and land-use impacts.

It is important to keep in mind that even manufacturing sectors that are thought to be pollution-free (for example, production of scientific instruments or photographic supplies) can cause serious environmental and land-use problems. Our work is thus equally applicable to those manufacturing sectors.

2. Quoted in David L. Brunner, Will Miller, and Nan Stockholm, *Corporations and the Environment: How Should Decisions Be Made* (Stanford, Calif.: Graduate School of Business, Stanford University, 1981), p. 135.

Executive Summary

In 1968 and 1969, Kelly Springfield, a subsidiary of Goodyear Tire and Rubber Company, took just 13 months from the time it began searching for a production facility site until it started turning out tires in Fayetteville, North Carolina. In December 1975, Standard Oil of Ohio (SOHIO) announced plans to build a major crude-oil terminal and pipeline at the port of Long Beach, California. The project, known as Pactex, needed about 60 environmental permits. Over three years later, in June 1979, SOHIO killed the project, citing, among other things, the environmental hurdles the project still had to surmount.

Clearly the rules of the game for siting industrial facilities had changed dramatically from the 1960s to the 1970s, and increasingly it seemed that plants were being blocked altogether by environmental regulations. Something important was in the air. Some in industry were claiming the United States had gone too far during the 1970s, when Congress and state and local governments enacted scores of new environmental and land-use laws. They alleged that such laws and regulations were stifling much-needed economic growth, eroding the competitive position of domestic industry, and causing long delays—and in some cases blockage—of new facilities set for construction.

Today, domestic economic problems have, if anything, brought environmental laws under even greater pressure. After Ronald Reagan took office in 1981, he made clear that he saw environmental regulations as a culprit in the nation's economic decline and an impediment to reversing it.

Obviously, the rules did change in the 1970s, when serious environmental problems throughout the United States prompted Congress and state and local governments to enact a host of new laws that firms had to consider when they started thinking about a new plant. Before 1970, building industrial plants—even big, heavily polluting ones like steel mills and petrochemical facilities—was fairly straightforward, with few uncertainties when it came to government regulations. Sites were chosen strictly from a business point of view, with corporate planners asking such questions as: How close are the markets? How good are the roads? How expensive is labor? Textbooks on industrial construction did not even mention environmental problems.

Industry's complaints notwithstanding, statistics reveal many impressive gains because of the new environmental laws—for example, dramatic improvement in air quality in many urban manufacturing centers. But, ironically, environmentalists are not entirely pleased with the new system either. They maintain that although on paper the system looks like it protects the environment, in truth the process doesn't always guarantee that projects are evaluated adequately or that the environment is protected.

Industry's "headline" concerns have given rise to several myths about the impact of environmental regulations. These myths have often obscured the real path to reform.

Myth No. 1: Environmental quality regulations cause industry to flee to other countries. The research for The Conservation Foundation's Industrial Siting Project failed to turn up any credible evidence that environmental regulations have precipitated, or are about to precipitate, a widespread exodus of American industry. In decisions about whether to build abroad or continue operating a facility in the United States, differentials in environmental-control costs are generally outweighed by production and other capital costs. Moreover, other traditional locational factors such as access to markets, proximity of supplies and natural resources, and political stability are almost always far more important than environmental regulations. At most, such regulations affect a decision only when all other factors are equal—which is rarely the case.

Only a handful of U.S. industries (notably, some primary-metal-processing industries and those producing highly toxic chemicals such as asbestos and benzidine dyes) have located more industrial facilities abroad in response to a combination of workplace-health standards, pollution controls, and a variety of nonenvironmental factors. These industries typically are experiencing static or reduced demand as a result of product obsolescence or hazards related to product use.

Myth No. 2: Environmental lures lead to interstate industrial flight. The issue of regional competition based on weak environmental laws and lax enforcement began nagging at Congress in the 1960s and led it to adopt national environmental standards so that industries could not play one state against another. The adoption of these federal laws quieted the debate over such "environmental blackmail" during the early 1970s, but later in the decade the issue flared again.

The concern was whether lax enforcement of these supposedly uniform regulations by some states might attract firms from other states. Again, the research for The Conservation Foundation's Industrial Siting Project said no: The right to pollute is not an important locational determinant. No evidence of a migration of industry from one state to another in search of "pollution havens" was unearthed, although there was a very slight tendency for states with lax environmental and land-use regulations to make relative gains in employment in pollution-intensive industries.

These states made even clearer gains in manufacturing as a whole, including industries unlikely to be affected by environmental regulations in making location decisions, which suggests that states with lax regulatory environments may also have other advantages (low labor costs, Sunbelt location, low unionization) that attract pollution-intensive industries. In fact, the most important aspect of environmental regulation may be its effect on perceptions about business climate. A state's reputation for a burdensome environmental-permitting system helps shape a company's notion of the state's overall business climate—often an important factor in locational decisions. By the same token, however, a reputation for strong environmental protection may increase a state's attractiveness to dynamic industrial sectors that attach great importance to quality of life.

Myth No. 3: Red tape is strangling industrial development. From a sampling of recent headlines, articles, and speeches, it might be thought that industrial projects no longer get built in the United States, thanks to environmental laws and other assorted regulatory ills. One study claimed that, between 1970 and 1978, half of all proposals for new or expanded oil refineries were blocked from locating at the site originally chosen and 58 percent of petrochemical projects were delayed because of environmental problems.

The research for The Conservation Foundation's Industrial Siting Project, however, contradicted this conventional wisdom: environmental and land-use regulations are not the primary cause of long delay in most industrial developments. In fact, a significant number of industrial facilities have been built relatively quickly over the past decade with few or no serious environmental problems. These success stories are often overlooked in the clamor over celebrated siting battles. Even in the headline-making disputes, delays caused by environmental quality regulations are often less significant than those attributable to financing problems, labor disputes, construction and equipment delivery snafus, lack of consensus regarding need, and regulatory hurdles not associated with environmental protection. When regulations do cause delay, that delay may be essential to protect legitimate public interests. Moreover, a good deal of the regulatory delay of the 1970s can be attributed to "teething pains" that are likely to be eased as the players in the siting game learn the new rules.

Myth No. 4: Other countries are better at reconciling environmental regulation with industrial development. Some industry officials argue that some of our allies—West Germany, France, Japan, and Canada—are more successful, for various reasons, at getting new projects built. Some say environmental laws abroad are less strict, while others maintain that governments and industries in Europe have eschewed the adversarial relationship prevalent in the United States for a more cooperative approach.

Once again, the Conservation Foundation research contradicted such popular wisdom: European countries such as West Germany and the United Kingdom have land-use controls far stronger than those in the United States, and it often takes a comparable amount of time to site a plant in Europe, Japan, or elsewhere as here. Furthermore, the lack of direct citizen participation in many siting decisions has helped fan bitter opposition. Siting disputes in Europe and Japan appear to be increasing and are no less intense than those in the United States.

Environmental conflicts overseas may become even more common as the European Economic Community adopts new and more stringent pollution control laws. In addition, European firms and governments may not be well positioned to deal with these problems; they are not as advanced as U.S. firms in assessing the environmental impacts of a project early in the planning process nor as adept at involving local citizens in the siting process.

The evidence that environmental and land-use regulations are not the impediments to industry that they are often portrayed to be is substantial. Yet there are compelling reasons to avoid complacency. The United States cannot ignore the significant costs that environmental protection entails, just as it cannot ignore the important benefits. Society needs to be concerned about the handful of industries that do appear to have suffered because of compliance costs. Furthermore, although economists generally agree that environmental restrictions do not have a large direct impact on the economy, they are still uncertain about indirect impacts on productivity, innovation, and inflation.

In some cases, businesses may decide against significant new investments because of the mere prospect of environmentally induced delays—what might be called "stillborn" projects. There is no real way to measure this phenomenon, although the theme recurs frequently in discussions with industry representatives.

Finally, both environmentalists and businesspeople ask whether the complex regulatory system actually provides the protections offered in theory by lengthy reviews.

When the industrial siting process is examined closely with an eye to identifying ways to remedy any problems while maintaining environmental-protection goals, four main weaknesses can be pinpointed:

- The regulatory system's confusing structure needlessly lengthens the permitting process.
- The system creates uncertainty, which plagues industry.
- The system does not always produce environmentally sound decisions.
- Decisions often lack finality, since administrative decisions can be reversed or challenged in court.

Concern with problems such as these is certainly not new. Much of the past focus has been on "cure-all" reforms to end problems quickly and neatly: one-stop permitting agencies, energy mobilization boards, relaxation of substantive provisions. Some measures would remove various governments, agencies, and laws from the process, while some would consolidate permitting agencies into a single body with responsibilities for all air, water, land-use, and other environmentally related programs and laws. Other proposals would restrict citizen participation, and several attempt to hog-tie regulatory agencies by imposing rules and procedures or by cutting personnel and budgets.

Effective, long-lasting reform, however, lies in an entirely different direction: "Quiet" reforms that focus on procedural and institutional changes are the best alternative. The current search by some for cure-alls may actually set things back, as promising, less-visible efforts are swept aside in the rush to reform.

Still, to solve many deficiencies in the system, government must reform. Lack of intergovernmental and interagency coordination, particularly in the review of large industrial projects, not only lengthens the review process but also often results in poor environmental-impact evaluation.

Federal, state, and local agencies are already experimenting with a host of techniques that seek to improve management with methods used every day in business to make organizations run more smoothly. Most operate without specific statutory authority and do not create new bureaucracies, add new regulations, or preempt existing laws. This approach has greater promise in the long run than consolidation and other cure-all reforms. Some examples of promising approaches already in use include:

- Designation of lead agencies to manage the permit process.
- Use of expert project teams within an agency to evaluate major proposals.
- Use of joint or consolidated hearings to reduce redundant meetings.
- Appointment of a single contact or project manager within an agency to keep track of how permits are progressing and to maintain contact with the project proponent; establishment of "escort services" within state governments to help identify needed permits, set up meetings with regulatory agencies, and monitor the permit process.
- Creation of mediating agencies to help settle interagency disputes over projects.

Even improved management techniques can be counterproductive, however, if they are not well thought out or not implemented with vigor or adequate resources. To some officials, "better coordination" has simply meant referring a proposal to more outside agencies and bodies, and then hoping for the best. Such informal referrals can be valuable, but they can also slow the permitting process without resulting in better reviews.

Governments at all levels can initiate quiet reforms that would improve the efficiency and effectiveness of the regulatory process. Efforts especially should be made to keep industrial projects from being delayed because companies cannot quickly learn what or when permits are needed, because regulations change midproject, or because regulators have insufficient time to handle all the projects assigned to them. Such problems can doom a project from the outset.

Some promising reforms are already being used in several jurisdictions and in agencies at all levels of government:
- Use of voluntary decision schedules to help keep reviews on track while fostering cooperation between regulators and project proponents.
- Reliance on project tracking systems to keep tabs on an application as it proceeds through the permit process.
- Careful use of grandfather clauses that protect industry from regulation changes once a project has started unless it can be shown that such changes are demanded by health concerns.

In addition, a few fundamental changes are needed to eliminate the dearth of properly trained government personnel overseeing the siting regulatory process. Experienced regulators should have incentives to stay in their positions. Their jobs must be given greater prestige, and they must be better paid.

Although the importance of government's cleaning up its own house cannot be underestimated, that step alone will not solve all the current problems in permit processing. In fact, the fallacy of many recent reform initiatives is that they look to government for all the answers, assuming that by tinkering with organizational charts or procedures all the problems will disappear. The need for a revamped company approach to planning and building industrial projects is of equal importance.

The more important changes for companies to adopt include:
- Earlier assessment of potential environmental impacts.
- Preparation of environmental reconnaissance statements that accompany each major capital proposal and that briefly outline potential environmental and land-use difficulties, how long they might delay project approvals, and how they might be avoided.
- More sophisticated internal organization for planning projects, including earlier involvement of environmental specialists and appointment of a single project manager to inform company executives of potential environmental and land-use problems and to deal with government regulators.
- A more open project-planning process, with early government and citizen participation.
- Greater attention to land-use and secondary-growth impacts of projects.
- Increasing use of alternative methods to settle siting disputes, such as mediation and mitigation.

Many of the problems in the regulatory system arise because the needs of corporate project planning do not always mesh well with regulators' needs. For example, if a state agency, with good reason, wants to know what air emissions can be expected from a multiplant project once it is fully developed, industry may respond that it can't be sure beyond the first plant or two, since changing markets might necessitate modifications in plant design.

Poor communication between industry and regulators is rife. Some of the fault lies with industry for not making clear at the outset its project schedule or the key decisional points in the project-planning process. Government must also bear much of the blame for this "communication gap," since regulators are particularly bad about letting applicants know where an application stands in the review process. The following cooperative steps by industry and government could give regulators badly needed information:

- Attempting to mesh industry's project planning with the regulatory process, rather than butting against each other. One option is to increase the use of preapplication meetings between company officials and regulators. Another possibility is the granting of outline approvals by regulatory agencies to indicate general agreement on the major elements of a project, with details to be worked out later. Such outline approvals are already widely used in local land-use regulatory processes for residential development.
- Adopting procedures to make necessary information available to regulators in ways that protect a company's trade secrets from the public and competitors.
- Using the so-called scoping process, whereby agencies from various levels of government get together early in the permitting process to identify key issues in a proposal and assign responsibility for studying them.
- Providing technical aid to local governments that lack expertise in siting concerns. This step should be taken by industry, as well as by federal and state agencies.

Similar steps to narrow the communication gap could help industry:

- Creating lists of sites that, for environmental reasons, are either off-limit or potentially acceptable for industries beginning a site-selection process.
- Compiling permit registers and guides that tell project pro-

ponents what permits they need, as well as when and how to get them.

Although the quiet reforms offered in this book involve far less disruption and change than some of the more radical initiatives being discussed around the United States, they still must be implemented thoughtfully and with great care.

Unless the proper foundation for change is laid, these reforms may never get off the ground, or, if they do, they may not work. A few basic precautions can help keep them from failing:
- Maintain the integrity of the system. The public and potential participants must have confidence in the regulatory process. If not, the likelihood that change will improve anything is diminished. Two ways to do this are to make sure that necessary checks and balances are in place and to involve all interested parties in designing the reforms.
- Provide adequate financial and administrative support. Regulators should not be burdened with new duties that cannot be met because funds or personnel are lacking.
- Give reforms high visibility, but do not politicize them so that they are discarded by a succeeding administration or management team.
- Take a comprehensive rather than a piecemeal approach. Comprehensive reform packages have a better chance of success than those that focus exclusively on one problem (as the early, one-stop permitting initiatives did).
- Be persistent. Don't expect perfection the first time around.

Although some of the proposals discussed here are far-ranging, there are limits to what can be accomplished through procedural and institutional reforms. Even if the system is made as efficient as conceivably possible, getting industrial projects built will still take time—one industry official estimated that it would take four to five years to plan and construct a petrochemical facility even if all regulations were repealed—and there will still be some rejections on environmental and land-use grounds.

Current regulatory reform efforts advanced in Congress and by the Reagan Administration do not hold a great deal of promise of getting at the real problems, which is unfortunate given the current national mood to make government more efficient. The downgrading of environmental-review offices in federal agencies, the continual bureaucrat-bashing that leaves regulators demoralized, the reduction in funds for resource surveys and for research to establish sound substantive standards, and the de-emphasis on enforcing environmental laws may well have political appeal, but they hold little hope of improving the process in the long run.

The continued popularity of environmental protection programs, as reflected in numerous public opinion polls, and the experience in Europe, where a grass-roots movement is forcing governments to adopt U.S.-style laws and standards, should make it clear that this country is unlikely to return to the old ways of siting industry. The adage about not being able to go home again rings true here.

The most promising methods of improving the efficiency of the environmental and land-use regulatory system are those that stress cooperation and negotiation. The innovative initiatives discussed in *Environmental Regulation of Industrial Plant Siting* offer relief without deforming the system, but they will work only if *all* the players in the siting game cooperate.

Chapter 1

Siting Industrial Facilities: The Rules Change

THE SITING PROCESS YESTERDAY: PARADISE LOST?

William Lever, founder of the massive English Unilever industrial firm, described how he chose the first plant site for his company:

> I went along the banks of the Mersey, on both sides of the river, taking trams to different points. Then I came across to where Port Sunlight is. I remember looking over the field gate . . . which led into the field in which the whole works now are. As I looked over and saw the water facilities and the railway, I said, "Here we are" and I never looked any further.[1]

This description prompts a nostalgic response: "Ah, for the good old days when life was simpler." But the good old days are gone. The consequences of unregulated industrial siting have been felt worldwide, and the United States is no exception. Serious pollution problems in the 1960s were apparent in every region of the United States, and a finger could often be pointed directly at lax environmental control of new industrial plants. By 1970, environmental and land-use regulations to curb pollution were being strengthened at all levels of government, a response that profoundly affects new plant sitings today.

Dramatic change over a century—between the time of Lever and today—is not particularly startling. But over a scant decade? Yet it was not until 1970 that environmental regulation on a grand scale became a force to be reckoned with. The dizzying speed with which environmental regulations were passed and implemented in the 1970s has left heads spinning—industrialists', environmentalists', and regulators' alike.

This torrent of environmental regulations has helped cause major changes in the industrial siting process—both in how industries behave when planning and building new facilities and in how regulators regulate. Understanding these changes is critical to evaluating environmental regulatory reforms that are now being bandied about.

Theory versus Practice

The way Lever chose his plant site hardly follows classic location theory, which holds that siting decisions are made by comparing economic costs like wage rates with benefits of possible locations such as proximity to markets. However, this theory is not always practiced in the real world. An expert in location theory pointed out: "The exact nature of the decision-making mechanism is an enigma. It varies from manager to manager, firm to firm, industry to industry and area to area."[2] As a business professor noted: "In the vast literature on corporate investment planning and decision-making, comparatively slight attention has been given to the actual behavior of firms in making decisions."[3] Thus, little help can be expected from siting theory in pinpointing environmental regulatory problems or in finding solutions to them.

And just as siting theory provides little guidance here, neither does the spate of statistics produced in recent years about the cost or inflationary impact of environmental regulations. These data may help gauge the macroeconomic impact of such regulations on industry (and almost everyone agrees the impact is not substantial), but they do not help identify regulatory glitches that should and can be remedied. Many studies about the regulatory process and its problems pay much attention to the "big picture"—the total cost of environmental regulations and the impact of those costs on productivity—but such a focus gives little counsel on real day-to-day problems in the system and on ways to solve them.

The contrast between the practices of the 1960s and the 1970s, on the other hand, graphically illustrates how many new rules have been written for the siting game and the problems that implementing and enforcing them cause today.

Siting in the 1960s[4]

Kelly Springfield

In late 1968, Kelly Springfield Tire Company, a wholly owned subsidiary of the Goodyear Tire and Rubber Company, began searching for a site to construct a new factory. Six months later ground was broken, and the first tire was produced by December 1969. The simplicity of the siting process in the 1960s is stunning when viewed through 1980s' glasses. Today it might take that long to get air-pollution permits.

Kelly Springfield decided to build a new plant after analyzing the geographic distribution of its markets and the location of its existing plants. There was a gap in the southeastern part of the United States, and its plants in Illinois and Maryland could not keep up with demand, necessitating shipment of tires from a Texas facility.

Based on this initial market analysis, the company undertook a preliminary study of freight costs for over 20 communities in the East and Southeast. Memphis, Tennessee, emerged as the least-cost location—the place where freight costs for assembling raw materials and distributing tires would be lowest—and Fayetteville, North Carolina, as the highest ($366,000 per year higher than Memphis).

In addition to transportation cost, perhaps the most important factor in the community search was the availability of labor; if the plant expanded as anticipated, it would require almost 2,000 employees. Thus, company planners felt that a labor supply of 150,000 people within a 30-mile radius of the plant was ideal; anything larger and the "labor climate" would be poorer (that is, perhaps unionized).

Because of the importance of a labor supply, the firm also wanted to avoid locating near existing tire plants that might compete for employees. In addition, since 40 or 50 supervisors would be required, the search team felt that the chosen community had to offer good living conditions to satisfy these key people. The ideal site, the team decided, would be an urban community with a minimum population of 25,000.

However, there were other considerations. The company foresaw increasing demand and ultimately wanted a plant large enough to take advantage of internal-scale economies. So a site able to accommodate rapid expansion, if the market demanded it, was sought. Availability of water, gas, and electric power was also a concern. Furthermore, the company hoped to hook into a municipal sewage-treatment system to avoid water pollution. Local and state taxes also had to be taken into account.

Guided by these factors, which had to be balanced against Memphis's freight-cost advantage, the search team narrowed the possible locations to 16. A small team (a production specialist, a personnel representative, and a plant planning engineer) then visited these locations for one day each and prepared preliminary field analyses.

Based on these analyses, Memphis's initial advantage evaporated when other locational factors were taken into account. The company settled on three cities—Greenville, Mississippi; Fayetteville,

North Carolina; and Johnson City, Tennessee—for comprehensive field analyses. Before these sites were scouted, local industrial developers were solicited to buy options on land at potential sites so that construction could begin as soon as a final choice was made. Several days were then spent in each city collecting details on industrial climate, labor climate, transportation facilities, economic costs, and living conditions. Some of the more important differences that emerged involved labor quality and training programs (Greenville lagged), quality of life (Fayetteville had a good number of colleges and many more attractive housing subdivisions), and site availability (there would be delays in preparing the site at Greenville).

Perhaps because the competing cities realized the clear advantage Fayetteville had in the firm's ratings (or perhaps because they did not), they engaged in a bidding war for the proposed facility, offering cheap sites, tax moratoriums, free water and sewer extensions, and improved highways. Fayetteville and the state of North Carolina ended up offering a variety of inducements, and that city was chosen for the new plant. The decision to locate was announced in February 1969. Construction began in April, and the first tire was produced by December.

Environmental and land-use regulations played only a minor role in this siting case.[5] North Carolina had not as yet imposed strong air-pollution control standards, and the Cape Fear River, where wastes would be discharged, was already classified for industrial use. Nor was local zoning a problem.

The major "environmental" issue turned out to be inadequate water supply. The production process Kelly Springfield planned to use needed a lot of water. However, during August, when the Cape Fear River is at low flow, its waters can be too salty for industrial use. So the city allowed the company to dilute the saline water with municipal tap water.

The ease Kelly Springfield had surmounting regulatory molehills is not surprising. There were not yet mandatory federal air-pollution standards, and federal water-pollution controls were weak. Land-use controls in North Carolina and most of the rest of the country were minimal. And environmental impact statements (EISs) were still just a glimmer in the eye of conservationists.

BASF

Not long after Kelly Springfield completed its siting process, changes began. The case of a chemical refinery, proposed by BASF, a major West German chemical company, for the coastline of South Carolina near Hilton Head, illustrates the changing mood and regulatory climate of the country.

BASF thought the site it had picked during the 1960s, in Beaufort County, South Carolina, for a $100-million petrochemical plant was a natural. The area was close to many petrochemical and dyestuff customers, and there were no other similar plants between Philadelphia and Miami. For two years in the late 1960s, the company had looked at more than 20 sites, mainly in Georgia, North Carolina, and South Carolina. In talking with state development officials, BASF laid out its basic needs: 1,800 acres of land, five million gallons of fresh water per day, proximity to a railroad and coast, a free-trade area to allow tax-free chemical imports, and an adequate, competitively priced labor force.

The state of South Carolina and Beaufort County were eager to please. The Beaufort area was racked by poverty. It was losing population. There were few manufacturing jobs, and even those paid poorly. If not for federal military installations, the area's economic picture would have been even more dismal. The state also wanted to maintain its record of attracting new investment, and it owned a piece of land in the Beaufort area that it had been trying to sell to industry for a decade.

By June 1969, BASF officials were convinced that Beaufort County was the place. In October 1969, the firm and state officials publicly announced the project, and, in November, BASF purchased the state-owned site along the Colleton River at a bargain price. Before all of this, BASF had sent, with few qualms, proposed plans to the South Carolina pollution-control authority. The company announced that all wastes dumped into the river would be treated. No state officials expressed concern about pollution, at least in public.

But there were rumblings of discontent. The developer of Hilton Head, a nearby resort community, first raised objections in June 1969. He argued that pollution from the plant would harm the tourist industry and cost jobs. The developer was later joined by a coalition of local citizens, headed by a retired Navy admiral, in fighting the project.

The coalition claimed that dredging for new docks would seriously damage the area's seafood industry. Coalition members pointed to loopholes in South Carolina's environmental-control laws that allowed BASF to pollute the area's waters. But what most incensed opponents was that BASF apparently planned to spend only $1 million on air- and water-pollution control. Other industry officials had allegedly told opponents that a budget of $7 to $8 million was more realistic if the most modern controls were to be used.

BASF was not to be thwarted. A majority of local people probably still favored the project. So the company promised to make this a model pollution-control plant, and, in January 1970, agreed to study pollution in the area with the help of the state water commission and an independent consultant. Meanwhile, construction plans would proceed on schedule.

Opposition to the plant intensified, however, when an investigation of a Tenneco Corporation plant by the state pollution-control authority disclosed that, despite representations by Tenneco to the contrary, the facility was polluting the stream into which its wastes were dumped.

Pollution control became a statewide issue. The authority, in its own defense, said that lack of funds and understaffing were to blame for its inability to monitor industry properly. Still, the authority and other officials spoke in favor of the BASF plant, seeking to assuage concerns about water pollution by pointing out that municipalities, not industry, were the biggest water polluters in the state.

South Carolina Senator Strom Thurmond then stepped into the fray. He asked the federal Department of the Interior to determine the real environmental effects of the BASF proposal. The state legislature also began to show some concern about pollution. Although turning down requests by the state pollution-control authority for more money, the legislature did give the agency more power internally and enacted a law to increase its enforcement authority.

But, in February 1970, local opponents filed suit; a second suit was filed by Hilton Head interests a month later. These developments troubled BASF, as did a March letter, from then-Interior Secretary Walter Hickel, warning BASF not to pollute the Colleton River estuary. Hickel relied, apparently, on an obscure provision of the federal Water Quality Act of 1965 that required special protection for pristine waters such as trout streams. State and local officials, outraged by what they saw as federal meddling, asked if South Carolina was to be "the guinea pig for a new federal policy on pollution."

The controversy simmered for the rest of the year. In January 1971,

BASF surprised everyone. The company pulled of the project, citing decline in demand related to the economic recession. Outsiders venture that BASF's financial problems, available capacity at other BASF U.S. chemical plants, and environmental opposition all contributed.

The environmental regulatory climate was changing. A senior federal cabinet officer had warned BASF not to pollute even *before* any of the major environmental legislation of the early 1970s had been passed (except for the National Environmental Policy Act [NEPA], which became law on January 1, 1970, a few months before Hickel's warning).

The change in climate caught corporations ill prepared. In a 1971 postmortem of the BASF battle, a business school professor from the University of South Carolina observed that: "It is almost inconceivable to the layman that a corporation the size of BASF would not have anticipated strong reactions from some segments of the population over the location of a possible polluter in the community."[6] Perhaps that judgment was too harsh. If a tire plant—not the cleanest facility in the world—could go into production so quickly, then siting a big petrochemical facility should likewise be relatively smooth, particularly when the project had strong state and local support.

However, the public mood was changing, and its will was being translated into a panoply of federal environmental laws. On the land-use side, states enacted legislation to protect coastal zones and other critical environmental areas. A 1972 *Business Week* article predicted that:

> All evidence indicates that even tougher fighting over land use lies ahead. It will test the ability of business to deal with shifting public attitudes as well as prescribed national values and priorities. At national and state levels, officials are inviting business to take part in designing new land-use controls. It will demand complex trade-offs between corporate economies and public interests.[7]

THE SITING PROCESS TODAY: OUT OF THE GARDEN OF EDEN

How had industry reacted by the mid-1970s to the flurry of regulatory actions? Were corporations considering environmental effects and land use in choosing sites for new facilities? Were they more sophisticated than BASF in dealing with environmental and land-use issues? A definitive business-school study on corporate environmental planning answered no.[8]

The study focused on project planning by 17 multinational corporations (MNCs) in three industries noted for having pollution problems—petroleum, chemicals, and metals. Twenty-one projects were assessed by interviewing, during 1974, 73 corporate executives. These interviews, supplemented by a survey of secondary literature and reviews of corporate documents, led to the conclusion that: "Project planning behavior in most MNCs is still unecological. Natural processes are often disregarded and full consideration of environmental values is typically absent. Project planners are not going through the intellectual exercise of asking 'what impact do our decisions have on the environment?' "[9]

Why were corporations so slow in responding to the public and legislative will? Are they any better today? It is essential, before addressing those issues, to first understand how the siting process works today.

Industry Perspective

The way new industrial plants are planned and built varies widely from company to company, given the great diversity of industries and differences even among firms within one industry. Figure 1.1, for example, illustrates the several steps that building a petrochemical facility might involve. There are, however, some common features in the manner in which firms make investment decisions, choose sites, and design individual projects.[10]

Determination of Need/Appropriation of Capital

The first basic step in building a new facility is to identify a capacity shortfall or an investment opportunity linked to market demand. Senior-level operating managers are responsible for pushing a project at this point. For example, a Texas-based chemical company that serves the West Coast market notes an increasing demand in the West. Two responses are possible: either expand in Texas and ship, or construct a new facility in the West to cut transportation costs significantly.

In some firms, identification of the need for new production capacity is ad hoc; in others, a very formal process, perhaps part of a multiyear plan, triggers such an evaluation. This preliminary internal review of the prospective investment typically includes a market analysis; determination of possible production and manufac-

Figure 1.1. Many Steps Lie on the Long Road between Project Proposal and Operating Plant

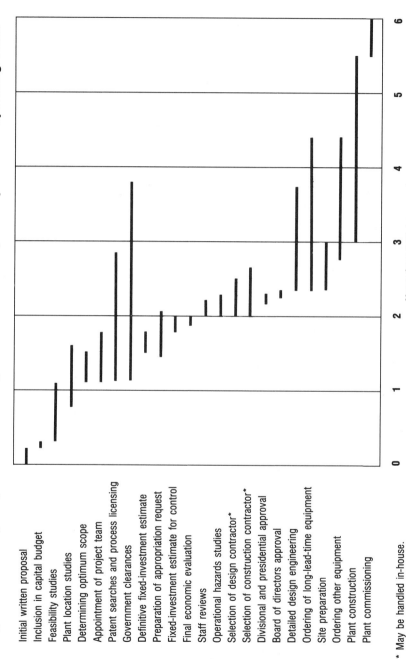

Reprinted with permission from *Chemical and Engineering News*, November 2, 1981, *61*, p. 40. ©1981, American Chemical Society.

turing process scale; a study of material, equipment, labor, and other inputs; and, finally, an overall long-term profitability analysis (often based on a net cash-flow analysis for the economic life of the project). In some cases, management may order an analysis to determine if expanding an existing facility is preferable to building a new one. This preliminary work, characteristically accompanied by an intracorporate debate over the proposed investment, leads to a formal request for capital appropriation or reservation by the corporate board.

Search for a Plant Site

Most proposals to expand or construct new plants never reach the second basic stage in building a new facility: looking for a site. Most corporations establish a site-search team, separate from any engineering/design work, for the proposed facility. Companies organize the team in a variety of ways to undertake the site search.[11] Some are highly centralized, directed and staffed from corporate headquarters; while others are decentralized, with regional divisions taking responsibilty.

A list of locational "musts" and "desires"—labor costs, proximity to markets and resources/supplies, and transportation costs—guides the team. Some industries have highly specialized locational needs to consider as well: aluminum refineries need cheap electricity; oil refineries require pipelines or deep-water access.

Most study teams first screen regions or states before zeroing in on particular sites. With this initial weeding done, the team then conducts a comprehensive, low-visibility investigation of a few sites, often fewer than ten, measuring each against the "must" list and evaluating comparative costs and benefits. Analytical information comes from a variety of sources: location consultants, state and local government agencies, or internally developed research. For the finalists that survive the screening process, a site visit to verify information and, as one observer puts it, "to get a feel for the place," is used.

At this point, particularly if economic considerations are equal, the company and state or local government begin horse trading. Negotiations might include tax concessions, labor training programs, or bargain-price sewage treatment for the plant. Studies to confirm site geology or lack of adverse environmental impact should also be under way.

Once a site is settled on, the board of directors typically commits to a full-blown project planning effort. At this point, the firm options land for the facility, perhaps with contingencies like rezoning approval.

Plant Design, Construction, and Operation

Once a site is chosen (and sometimes before), a separate project team, different from the site-search team, is assembled to oversee detailed project planning: engineering and design, public relations, equipment ordering, and the securing of government approvals. The project team's composition may vary markedly from project to project and firm to firm, but the team generally includes a leader, engineering and design specialists, government relations people, and outside consultants. Very early in this stage, the team seeks necessary government approvals, including environmental and land-use permits.

Once the firm senses that the project will get a green light from the various governments involved, and if the economics still make sense, it begins selecting contractors, drafting detailed design and engineering plans, and ordering equipment with long lead times. Construction generally waits until necessary approvals are granted.

Such a siting process looks logical and certain on paper. Of course, real life confuses the logic. Some firms have little flexibility in picking plant sites—after all, a mining company must go to the minerals. Others may select a new location based on the personal preference of their chief executive officer. Or economic uncertainties—business cycle fluctuations or unpredictable inflation rates—may torpedo the most well-conceived project.

Neither is the planning process clear-cut. Its stages often overlap; work on government approvals may begin before the board of directors places its imprimatur on the project. However, regardless of the permutations in the planning process, the mark of environmental and land-use regulations on this three-step process is often indelible.

Regulatory Perspective

The steps that industries must follow to secure environmental and land-use permits from regulators also overlap. In addition, they vary from project to project and place to place. Small industrial facilities typically need few federal permits or state and local land-use approvals. In contrast, larger projects contend with both federal and state pollution-control regulation, along with state siting laws and local land-use controls. There are, however, three basic stages in permit acquisition.[12]

Preapplication Permit Identification

The first step in the permitting process, identifying necessary environmental and land-use permits, is the culmination of the firm's summary analysis of environmental constraints associated with the site. This analysis may begin early in the site selection process, but more likely it is not started until the company narrows its search to a few sites or even commits itself to a single location.

Most major new industrial facilities need permits under the Clean Air Act and the Clean Water Act, issued by the U.S. Environmental Protection Agency (EPA) or state agencies that have been designated to administer federal programs. If dredging and filling in a navigable waterway—defined very broadly to cover most wetlands and water courses—are anticipated, then a dredge-and-fill permit is needed from the U.S. Army Corps of Engineers. Special federal EPA permits for the disposal of hazardous wastes from a firm's production process may also be needed.

If the necessary permits or approvals are from any federal agency other than the EPA, and the project will significantly affect the environment, then in most cases that agency must prepare a detailed environmental impact statement (EIS) under NEPA. If a pipeline right-of-way over federal land is necessary, chances are an EIS will be required for the entire project.

Over half the states have environmental policy acts that also require impact statements where state approvals are involved, although the scope of state "little NEPAs" varies widely. A few are strict, forbidding issuance of state and local permits if a project will have serious, and unavoidable, adverse environmental effects. Others are weak, providing few procedural or substantive protections.

One document often can now satisfy both federal and state requirements for environmental impact statements, although this was not generally the case until 1977. However, several states require some information in addition to EIS before agencies can approve major industrial facilities, particularly those proposed along coastal areas or in or near critical areas, such as estuaries, marshes, scenic landscapes, or prime agricultural lands.

Local regulation focuses mostly on land-use and indirect growth impacts of a facility—traffic, housing, and sewage-treatment requirements. Some local governments do have special restrictions on industrial development in sensitive ecological areas or on construction of hazardous waste facilities, although many have no zoning requirements at all. Perhaps the most common local permission necessary is rezoning, which often provides leverage for a community to control non-land-use impacts of a project.

Necessary permits are typically identified by informal preapplication meetings with regulatory officials at all governmental levels. During these preapplication meetings, firms try to get a sense of whether the project can navigate the permit system, but regulators usually are reluctant to make any commitments without seeing more detailed engineering drawings or environmental impact analyses. Even if red flags are raised by regulators at this point, however, other aspects of project planning can proceed.

Preparation of Permit Applications

Before preparing applications, particularly for federal air permits, firms today must often monitor the proposed site to determine existing air quality and gather general ecological data. Government sources in some areas of the United States collect such data, but most often the firm must develop the information. This work, usually contracted out to consultants, rarely starts until a firm feels confident of ultimate approval.

If a firm senses approval, it may also begin working (again, usually through consultants) on a full-blown EIS if one will eventually be required. This stage is also the time to investigate potential mitigation of unavoidable adverse environmental impacts. For example, if a project involves transport of oil to a site, a company might begin developing a spill-control plan.

Submission and Processing of Applications

What appears to the casual observer to be the beginning of the siting process—the submission of permit applications—is really the tail end of the plant location search. Only detailed design work and actual construction remain. By this time the firm is usually wedded to a specific site and feels confident it can garner all needed government approvals.

Most industries submit applications concurrently to federal, state, and local agencies. Thus, a company may have air permits pending with the federal EPA, water permits with a state water-pollution board, and a rezoning application with a local authority. While permits are pending, the submitting firm may help an agency prepare a detailed environmental impact analysis.

In many cases, agencies return applications to companies for more information. Most companies disclose no more information than necessary in the permitting process. They fear inadvertent disclosure of important trade secrets to competitors, or they worry that providing too much information to regulators makes them curious about project details unrelated to the permit application.

Of course, regulators want as much information as possible. They want to know exact production processes, pollution-control equipment, operating schedules, waste handling techniques, and anticipated number of personnel. Companies often reply that it is too expensive to develop specific information until the project has at least tentative approval. This stalemate engenders a good deal of haggling over what information the company must produce.

Contrary to popular belief, the actual evaluation of permit requests is not a ministerial act, requiring only the application of clear, preestablished standards. Even where there are standards to guide issuance of an air permit, leeway always exists because measurement and modeling techniques[13] are flexible. Likewise, the "best available" control technology, required by the Clean Air Act, is negotiable. On the state and local levels, authorities have great flexibility in granting rezoning or development permits based on public health, safety, and welfare.

Industry is not without its bargaining chips. Political pressure can be brought to bear on regulators. In fact, in some instances final decisions are made by elected officials; their actions are accorded great deference by the courts. Thus, even if professional land-use planners want to deny a permit to rezone prime agricultural land, they can be overruled by the local elected council.

Characteristically, regulators see the application process as the first step in enforcement. By scrutinizing a project and imposing design controls, air-pollution regulators have less trouble enforcing permit conditions later than if the industry is given free rein to meet a particular standard. Similarly, a land-use permit may require alteration of a project to avoid the *possibility* of adversely affecting a nearby wetland.

Regulators know from experience that it is easier for them to prevent an adverse impact than to stop it once a plant is in operation. The legal process can be slow and cumbersome, and once people are hired and at work it is difficult to shut down a facility while adjustments are being made.

Once an application is complete, hearings are usually required before a permit can be issued. Multiple hearings are frequently necessary because of fragmented jurisdiction over projects: a firm may find itself before a state air-control board one week and a local zoning board the next. For some permits, particularly federal ones, the responsible agency must solicit comments from other agencies, levels of government, and the public.

After comments are obtained and hearings held, the permit is issued if statutory criteria and standards are met. Challenges to a permit or denial thereof, either by industry or project opponents, generally must be filed within a fixed period—30 to 60 days is typical—after issuance.

CONCLUSION

In the 1980s, industrial plant siting and attempts to reform that process require a sophisticated understanding of new environmental and land-use regulations. But it is not enough, for either an industrialist or a regulator, to know only how this new process affects one's own role in siting; each side of the issue must be examined. Regulators need to remember that, for industry, siting is a high-stakes venture. Industry officials need to understand the legal and political constraints facing regulators. Such understanding is essential to make the process work better. And all parties involved in industrial siting need a better picture of just how seriously the new rules have affected the building of new facilities:

- How do environmental and land-use regulations affect the siting of individual industrial facilities? Are they so onerous that U.S. firms might actually forgo new capital investments or close up shop and go to other states, perhaps even to other countries, with more favorable attitudes and laws?
- Are many industrial projects being strangled by lengthy environmental reviews?
- Are there ways to improve the efficiency *and* effectiveness of the environmental review process for industrial plants?

These questions are considered in the chapters that follow.

REFERENCES

1. Charles Wilson, *The History of Unilever*, quoted in C. S. Deverell, *The Location of Industry* (Sussex, England: Editype Ltd., 1968), p. 9.

2. Howard Stafford, *Principles of Industrial Facility Location* (Atlanta: Conway, 1980), p. 252.

3. Thomas Gladwin, *Environment, Planning and the Multinational Corporation* (Greenwich, Conn.: JAI Press, 1975), p. 11.

4. The two case histories related here are highly summarized. The Kelly Springfield case comes from an excellent recent book on industrial location: Barry Moriarty, *Industrial Location and Community Development* (Chapel Hill, N.C.: University of North Carolina Press, 1980), p. 121. It was supplemented by a telephone interview with Moriarty.

The BASF summary is from a detailed case study by a team of business professors: Oliver Wood et al., *The BASF Controversy: Employment vs. Environment* (Charleston, S.C.: College of Business Administration, University of South Carolina, 1971). Another useful source of case studies, both on a firm-by-firm and an industry-wide basis, is David M. Smith, *Industrial Location* (New York: John Wiley & Sons, 1981).

5. Telephone interview with Barry Moriarty, February 5, 1982.

6. Wood et al., *The BASF Controversy*, p. 66.

7. "The Land-Use Battles Industry Faces," *Business Week*, August 26, 1972, p. 40.

8. Gladwin, *Environment, Planning and the Multinational Corporation*, pp. 232-34.

9. Ibid., pp. 256-57.

10. An excellent short article on industrial-site selection, with emphasis on environmental factors, is James Cleveland et al., "Using Computers for Site Selection," *Environmental Science and Technology*, vol. 13, no. 7 (July 1979), p. 792. For in-depth discussions and analyses of industrial-location decisions, see Stafford, *Principles of Industrial Facility Location*; Smith, *Industrial Location*; Moriarty, *Industrial Location and Community Development*; Roger W. Schmenner, *Making Business Location Decisions* (Englewood Cliffs, N.J.: Prentice-Hall, 1982). The Society of Industrial Realtors has published several useful references, including a siting checklist, *A Guide to Industrial Site Selection* (Washington, D.C.: Society of Industrial Realtors and the National Association of Industrial and Office Parks, 1979); and an extensive bibliography, *Industrial Real Estate: An Annotated Bibliography* (Washington, D.C.: Society of Industrial Realtors, 1982). Other useful sources are two bibliographies by James E. Rowe: *Industrial Plant Location* (Monticello, Ill.: Vance Bibliographies, 1980), and *The Theory of Industrial Location* (Monticello, Ill.: Vance Bibliographies, 1980).

11. For an excellent discussion of this point, see Schmenner, *Making Business Location Decisions*.

12. For those interested in a detailed, but somewhat dated, description of federal environmental law, see John Quarles, *Federal Regulation of New Industrial Plants* (Washington, D.C.: published by the author, 1979). A comprehensive discussion of state land-use programs can be found in Robert Healy and John S. Rosenberg, *Land Use and the States* (Baltimore, Md.: Johns Hopkins University Press, 1979).

13. Air-quality modeling is a mathematical technique for estimating the effect of an air-pollution source on air quality at various locations. Computers are often used in the modeling process.

Chapter 2

Environmental Laws: Roadblocks to Economic Growth?

The siting process for industrial facilities, particularly large, pollution-intensive ones, underwent a dramatic metamorphosis during the 1970s. The environmental and land-use regulatory system did not develop overnight, however; it evolved over a decade. As a result, industry faced a great deal of uncertainty as new laws were enacted and implemented. Regulators also found the period unsettling as they tried to administer a host of new requirements, legislated discretely, that were often compartmentalized and uncoordinated.

In general, the regulatory process became more complex with a great deal more discretion over project approvals vested in government officials. Not surprisingly, players on both sides in the siting game soon began to detect problems, such as those illustrated in the case studies in chapter 1, that cried out for reform. These concerns began to make headlines and today continue to worry industry officials, government representatives, and environmentalists alike.

Neither government regulators, environmentalists, nor the public have a feel for how significantly the process changed in a short time. Nor do they fully understand the steps involved in bringing an industrial plant on-line in the 1980s, particularly the early stages of project planning before submission of permit applications. It is then that the regulatory process kicks in and, with increasing frequency, that trouble arises.

Yet it is evident time and again that crucial decisions about a project often need to be made several years before permitting and that changing plans in midstream is difficult and time-consuming. By the same token, there is a disturbing lack of understanding among industry officials, even those involved in plant siting, about how the environmental regulatory process works and about the demands on and needs of government regulators.

This apparent lack of understanding leads to a host of unanswered questions: Will environmental and land-use laws be given short shrift in the present drive to sustain economic recovery? Will environmental quality be sacrificed in the longer term effort to make the American economy more competitive in world markets? These are not just rhetorical questions. The fundamental issue is whether such laws and environmental quality *need* to be sacrificed. Are they really so burdensome, so complex, that nothing short of radical surgery will cure the regulatory ills? Or are industry and regulators just experiencing "teething pains" that can be remedied without drastic changes?

CELEBRATED SITING DISPUTES[1]

Before answering questions such as these, it is helpful to focus on where things have gone wrong. Any shortcomings in the regulatory process should show up in a study of celebrated industrial-siting disputes like the cases that follow. It must be kept in mind, however, that these cases are not the norm but the worst of their kind. For every horror story, there have been many siting successes where the system worked, delays were few, and plants got built. Those success stories form the basis for suggestions made in chapter 6 on how to improve the regulatory process.

Dow versus California[2]

The Project

In early 1975, the Dow Chemical Company unveiled plans to build a $500 million, 13-unit chemical production facility on the banks of the Sacramento River, 35 miles northeast of San Francisco. Half the complex would be built on a 2,700-acre parcel in undeveloped Solano County. The rest of the complex would be constructed in the industrialized town of Pittsburg, California, just south and across the river in Contra Costa County, where Dow had operated a chemical plant since 1940. Submerged pipelines would connect the two sites. The new facility, designed to produce basic chemical building blocks, such as styrene, vinyl chloride, and ethylene, would employ 1,000 construction laborers temporarily and 800 permanent workers when operational.

Dow hoped to take advantage of relatively cheap and plentiful Alaskan oil that would become available in the late 1970s. In addition, the company would save $56 million annually on the cost of shipping products from the Gulf Coast states to the West Coast. With these savings, Dow could reduce its prices and capture a larger share of the huge California chemical market, which was served primarily by plants east of the Rockies.

From a business point of view, this site made much sense. Because it already operated in the area, Dow would not have to import key management and technical people. Moreover, the company knew the local area and its politics. Most important, the Solano location was one of the last deep-water sites on the West Coast that could accommodate large tankers. Perhaps because this location seemed so ideal, Dow did not consider many alternative sites. It toyed for a while with simply expanding its existing Pittsburg facility, but the company could not acquire necessary adjacent land. Dow also looked at sites near Los Angeles and at one in Washington State. But, all in all, the Solano site looked best.

Because its production facilities in Texas and Louisiana would be fully committed to supplying other markets by the early 1980s, Dow felt it imperative to have the Solano facility completed by 1982. It served notice on the state and local government that speed in approving the project was essential. Company officials thought this demand reasonable since, according to Western Division personnel, no red flags had been raised in early meetings with government officials and conservationists in 1974.

The Western Division had not built a new plant for nearly 40 years. The siting team—an engineer, environmental specialists, and government relations personnel—that the division organized was assisted by an experienced state environmental consultant and a well-known local attorney. This team answered only to the Western Division's head, who had conceived of and promoted the project at Dow's headquarters in Midland, Michigan. Once Midland appropriated capital for the project, the Western Division was on its own.

The Economic, Social, and Political Context

Unemployment in California stood at a record 9.4 percent as the recession of 1974-75 deepened. The governor, Jerry Brown, found himself under fire from business and labor for his support of environmental initiatives. Skilled laborers and construction workers in the Bay Area brought pressure on local politicians to create jobs. Solano County, a small but fast-growing, conservative Bay Area jurisdiction, which had heretofore relied largely on agriculture and military bases for its economic strength, rolled out the red carpet for big industry in hopes of increasing its tax base. Solano County calculated that Dow's project would add 14 percent to its assessed valuation without bringing in many new residents who would demand costly services.

The Environmental Regulatory Process

Dow needed 65 permits and approvals for the project, but many were routine—building permits and the like. Several key approvals had to come from Solano County: rezoning of the prospective site from agricultural to industrial; cancellation of an open-space preservation designation; and certification of a state-required environmental impact report (EIR) that the county would take the lead in writing. Other hurdles included obtaining air-emission permits from the regional air board based in San Francisco, consent from various state agencies to withdraw river water and run pipelines across state property, and a dredge-and-fill permit from the U.S. Army Corps of Engineers to construct a turning basin for ships. This last permit required a federal EIS, as distinct from the state EIR.

Environmentalists, mainly from nearby Berkeley and San Francisco, had deep concerns about the project's impacts. The proposed site was adjacent to the Suisun Marsh, a key feeding and breeding ground for birds along the Pacific Flyway and a spawning ground for a locally important salmon run. Any spills from oil tankers servicing the new facility would wash into the marsh. But the Dow EIR paid little attention to this issue.

Opponents also worried that cancellation of the local open-space designation and rezoning from agricultural to industrial would set a precedent for further heavy industrial development in the area, then largely farmland. California's state agricultural land and open-space protection law (known as the Williamson Act) was new, and this dispute was viewed as a test case of the law's effectiveness.

Air quality in the area already fell below federal and state standards, and evidence suggested that air pollution from the complex would be transported to the San Joaquin Valley, one of California's most productive agricultural areas. Moreover, water withdrawal would increase the intrusion of salt water upstream and affect the area's importance as a spawning ground.

The Outcome

Dow succeeded in getting all necessary approvals from Solano County, including certification of the EIR, in near-record time. Other state and local agencies expedited their reviews and approvals of the EIR, and the company appeared poised for success. It looked like Dow would break all records for getting a big project approved in California.

Things were not to be that easy, however, and soon Dow had its back to the wall. Environmentalists filed suit challenging certification, the rezoning, and open-space dedesignation. Not long thereafter, the regional air board denied Dow an air-pollution permit on the grounds that the area's air already violated federal ambient standards. Because there was no national "offset" policy in effect, the regional air board said it could not give Dow credit for cleaning up its existing facilities to offset pollution from the new complex.[3]

State agencies, which had already "signed off" on the EIR, kept asking for more data, holding up permit approvals. The state agencies worked to delay federal EIS approval as well. Governor Brown finally called for a consolidated hearing on Dow's applications for state permits and directed state agencies to submit a final list of questions about the project. The hearing turned out to be a fiasco from Dow's point of view, ending with the prospect that the EIR would be completely reworked. Even then, however, there would be no guarantee of approval from the state agencies or the regional air board.

The Western Division was competing with other national and international branches of Dow for a chunk of the company's capital budget. When the project stalled without any clear path to regulatory approval, the board of directors reallocated funds to projects that looked more promising. In early 1977, the company canceled the project, blaming the state for bureaucratic delays and environmentalists for obstructionist tactics. "The system didn't decide against the plant," quipped one Dow official. "It confused it to death."

Reaction across the state was swift. Business and labor leaders were outraged. Corporate executives told their brethren in other states to use the "ABC" theory of plant location—"Anywhere But California." Governor Brown supported legislation to cut environmental "red tape" at the state level and sported "California Means Business" buttons to lure jobs to the Golden State. Environmentalists were put on the defensive, and some say their reputation was permanently damaged in the public's mind.

Dow's difficulties resulted in some positive regulatory action, however. At the national level, the Environmental Protection Agency (EPA) adopted an offset policy, later ratified by Congress, that relieved some of the pressure to weaken the Clean Air Act. The state of California organized an Office of Permit Assistance to speed the environmental review process without changing the substance of any laws, and the Williamson Act was amended to clarify key provisions. Even Solano County improved its land-use planning process to assess more adequately the impact of major industrial development. Many people believe the project could win approval today, but Dow has no plans at present to build, although it retains ownership of the land.

While regulatory difficulties clearly contributed to the demise of the Dow project, some critics claimed that headquarters pulled the rug out from under the Western Division too quickly. Others alleged that with the possibility of a quick approval fading, the project made less sense because Alaskan oil supplies would already be declining by the time the facility was on line.

What was the primary reason for the delays? In reality, a combination of problems in the siting process contributed to Dow's decision to pull out. Some were self-inflicted wounds that Dow could have avoided; others were directly attributable to a squeaky regulatory machine.

Analysis of Regulatory Difficulties

Dow brought a good many regulatory problems on itself by rushing the project in its early stages. While this haste is understandable given the intense intracorporate battle for funds, Dow demanded approvals in a time frame that, in California, was optimistic for even a small housing developer.

The Western Divison did evaluate alternative sites, but none was considered seriously, except for the possibility of expanding the existing facility. Environmental or land-use considerations apparently played no significant role in evaluating possible sites. If they had, perhaps Dow would not have paid so much money for a site with serious land-use restrictions.

However, Dow did meet with government officials and conservationists before announcing the Solano County project site and, based on those meetings, felt there were not any insurmountable problems. After these preliminary meetings, Dow pushed hard for quick approvals. At the county level, the company got them in near-record time.

But the seeds of destruction had been planted. Solano County was only too willing to oblige Dow by quickly rezoning Dow's property from agricultural to industrial, even though the comprehensive land-use plan for the area was unclear about future development. Solano County also quickly canceled the agricultural land/open-space designation made pursuant to a statewide law, thus establishing a precedent-setting case that had implications for open-space and agricultural land protection throughout the state.

Finally, Dow pushed the county to approve the state-required EIR quickly, which it did. Much of the information for the EIR came from a consultant paid for by and directly responsible to Dow. While that consultant had a good reputation, the company pressured for quick action; moreover, the appearance of Dow's possible influence over the consultant's objectivity gave ammunition to opponents.

Besides opening the company to later legal challenges, Dow's haste also raised hackles among state agency regulators. They were given only a week or so to evalute the project by the time Solano County circulated the draft EIR, as required by state law. Only later, when Dow had to go back to those agencies for permits, did the agencies request more detailed information for particular permit applications.

Haste may have also contributed to substantive shortcomings in the EIR, which did not cover some crucial environmental concerns. Information on oil spill hazards, for example, had to be developed later at the insistence of project opponents.

There were also other more subtle gaffes. Dow did not make it clear that the facility might produce vinyl chloride, a toxic substance linked to liver cancer. When opponents discovered that possibility, Dow appeared to be covering up.

Dow is criticized for the way it dealt with state agencies. A frequent complaint is that Dow personnel and the attorney the company hired treated state bureaucrats with disdain. Rather than negotiating over points of contention, critics say that instead Dow adopted a pugnacious approach.

Local governments had a hand in the outcome of Dow's debacle. Those involved all favored the project. In fact, they wore such feelings on their sleeves and repeatedly bent over backwards to expedite approval. This expeditious treatment smacked of favoritism and backfired on Dow.

Solano County was not well equipped to evaluate the impact of a $500-million petrochemical facility—neither the direct environmental impacts nor the long-term indirect growth that the project would stimulate. Not that the county was without resources; it did have a small planning department that, while not highly sophisticated, was better staffed than many other local jurisdictions across the country. Overall, however, its resources were overtaxed, a situation made worse by the pressure Dow exerted.

State agencies gave Dow the biggest headaches, at least in the company's view. In theory, the various agencies had an opportunity to comment on and flag any problems with the proposed project when the EIR was circulated. Unfortunately, most did not because of time and personnel limitations, given Dow's desire to expedite approval. The agencies may have figured on getting another crack at Dow when the company had to come back for specific permits.

Actually, most of the permits Dow needed from the state agencies were minor—the crucial environmental ones involved withdrawal of water from and discharge of effluents into the Sacramento River. But people in other agencies who had questions—such as in the state air-resources board and the agriculture department—saw an opportunity to jump back into the fray.

A few state agencies quickly issued permits (some later said they had approved the project without much review because they simply did not have the expertise to evaluate it). Other agencies with quite limited jurisdictions started asking detailed questions about impacts that had nothing to do with any permits they might issue. The theory was that the California Environmental Quality Act required any agency to consider *all* environmental consequences, not just those related to its own permits.

To remedy this situation, the Governor's Office of Planning and Research offered to hold consolidated hearings to get all questions on the table once and for all. Dow refused initially (some say because of a divide-and-conquer strategy), but later accepted. More and more questions piled up. Finding answers would cost more time and money. Dow killed the project.

Later, the California legislature tried to remedy the situation by changing the state EIR law. Agencies' jurisdictions were restricted, and questions had to apply only to the permits before them. Thus, in granting a water-effluent discharge permit, the state water-resources-control board cannot now consider air-quality issues.

Dow's biggest pollution-control obstacle was the regional Bay Area Air Pollution Control District (BAAPCD). BAAPCD, an independent agency, had regulatory authority over all industrial plants in the region. Neither the federal EPA nor the state air-resources board had direct control over the BAAPCD. While Dow's first plant to go on line would itself meet all emission standards, the BAAPCD's interpretation of the Clean Air Act and district regulations did not permit Dow to build, even if the company cleaned up its existing plants or others in the vicinity of the proposed facility, because San Francisco's ambient air already exceeded federal standards. Both the federal EPA and the state air board felt that use of such "offsets" was permissible, but BAAPCD refused to budge and denied the necessary permits. BAAPCD told Dow that Congress must amend the federal Clean Air Act before it would allow offsets—which Congress did in 1977.

Several federal agencies had direct roles in the Dow project, notably the Army Corps of Engineers, but they generally stayed on the sidelines. The Corps did begin work on a federal EIS, but stopped at the request of the state resources agency head, who said that available information was inadequate.

The regional office of the EPA did make known its belief that the regional air-pollution-control district could have adopted an offset policy, but otherwise it played a minor role. Statutorily, it had no authority over the project since it had delegated regulatory power to California (as permitted by the Clean Air Act upon a finding by the U.S. EPA that a state program is acceptable). When Dow canceled the project, the EPA issued a press release making clear that it had nothing to do with the demise.

Environmental groups' opposition probably slowed permit approvals, particularly at the state level. However, the threat of drawn-out litigation was the opponents' biggest stick and, according to Dow, was a major reason the project was killed. Dow thought it might get all the permits, yet still be tied up for years in court.

Later developments, notably a California Supreme Court decision in an unrelated case, showed that the environmentalists' suit, challenging various local approvals, probably would have succeeded on the agricultural land/open-space designation issue alone. If the agricultural designation of Dow's property had remained, it would have delayed the project for up to ten years. Later, however, the California state legislature enacted an escape clause that Dow could have used to void the agricultural land protection contract.

SOHIO's Pactex Project[4]

Just as the Dow project was being canceled, the Standard Oil Company of Ohio (SOHIO) was also announcing an industrial development in California that would make headlines. The cast of characters differed somewhat at the state level, although some of the issues involved—notably those regarding air pollution—were almost identical. Had California learned anything? Would SOHIO be more adroit than Dow had been?

The Project

In December 1975, SOHIO announced plans to build a $500 million crude oil terminal and pipeline at the port of Long Beach, California. The project, named Pactex, involved marine transport of oil from Valdez, Alaska, to a marine terminal at Long Beach, where eight tanks with a 700,000- (later changed to 500,000-) barrel-per-day capacity would provide temporary storage.

Dredging would be required to accommodate large tankers, and newly filled-in areas would house three of the storage tanks on Pier J, the terminal site. Oil would be piped to Texas, entailing construction of over 200 miles of pipeline and conversion of 800 miles of existing natural gas pipeline running westward into an oil pipeline flowing eastward. From Texas, the pipeline would hook up with an existing oil pipeline network.

At first, benefits of the project to California appeared to be few. The operations, designed to supply markets in the Midwest and East with crude oil, would employ 550 construction workers temporarily and 80 persons permanently. The project offered minimal tax revenues to anyone except the port, a virtually autonomous entity with no schools or residences to support. On the corporate hand, SOHIO would avoid transporting oil from Valdez through the Panama Canal to Louisiana. World market prices, combined with federal oil price controls, meant that transportation costs were absorbed by the company. The Pactex project would eventually save SOHIO between $.50 and $1.75 per barrel compared with the alternative of Panama Canal transport.[5]

To SOHIO, the port at Long Beach was a natural site. While locating anything on the California coast presented major difficulties in the wake of a 1972 initiative that created an environmentally conscious state coastal commission, both the initiative and the newly developed state coastal plan favored the development of existing port facilities, as opposed to less intensively developed or utilized areas. The port was a heavily developed area, with oil destined for local refineries already providing much of its revenue. In fact, some environmental leaders encouraged SOHIO to site in Long Beach, figuring they were not losing anything worth saving anyway.

Time was a most important advantage of the Long Beach-to-Midland, Texas, project. SOHIO calculated a two-year savings over alternative northern pipeline routes. And from an engineering point of view, the project posed few problems. The job could be completed within two years. With the trans-Alaskan pipeline (TAPs) scheduled to produce oil by mid-1977, any west-to-east transportation project had a defined time limit.

The Environmental Regulatory Process

SOHIO, a regional refining and marketing company, had undergone a major transformation in the late 1960s. It negotiated a deal with British Petroleum (BP) that made BP the majority common stockholder of SOHIO. In return, SOHIO received all of BP's rights to North Slope Alaskan oil, and, as a result, immediately became one of the largest owners of domestic oil reserves. At the crucial early stages of Pactex, many of SOHIO's top-level management and engineering team continued working on the TAPs project in Alaska. Consequently, the company's mid-level staff coordinated permit activities; management at corporate headquarters in Cleveland retained all major decision-making powers.

Before SOHIO chose the Long Beach site, company representatives met with state officials and environmental leaders to sound them out about the regulatory process. No one knew much about the air-quality implications of the project. Dredging and water permits seemed the biggest hurdles. Overall, the regulators seemed guardedly optimistic, but they were reluctant to make any commitments until project details were fully known.

An astronomical 703 permits, government approvals, and private rights-of-way agreements were required of SOHIO. But of these, only 89 involved discretionary decision-making; the rest were ministerial or routine. Fifty-nine of the discretionary permits were environmentally related, including various air, water, and coastal-zone approvals in addition to a state EIR and a federal EIS. The other 30 involved construction requirements.

The project's scope is best understood by looking at the number of jurisdictions associated with the permits. Permitting spanned all levels of government and included 10 federal agencies, 22 state agencies, 12 counties and 22 county agencies, 19 cities and 52 municipal agencies, and 6 special districts. Unlike the TAPs project, which involved only Alaska, Pactex crossed four states, requiring approval in each. Additionally, SOHIO had to secure permits or approvals from four railroads and from four corporations or individuals. To accomplish this enormous task, SOHIO established an affiliate company on the West Coast, the SOHIO Transportation Company.

Environmental Concerns

As project details unfolded, the League of Women Voters, the city of Los Angeles, a coalition of Long Beach citizens (the Citizens Task Force on SOHIO), and some state officials began voicing significant concerns about the environmental impacts of the project:

- *Air Pollution.* The port of Long Beach is located in the South Coast (Los Angeles) air basin, known throughout the world for having poor air quality. Added air emissions would come from tankers, vapors from storage tanks, increased electrical generating power from nearby power plants to pump the oil; moreover, the abandonment of the natural gas pipelines meant that Los Angeles industries would have to rely on "dirtier" sources of power. Offsets would be required so that the project, in effect, would result in cleaner air for the area.
- *Coastal Zone Impacts.* Three large oil storage tanks were to be constructed at the terminal site. State coastal plans approved by the federal government required that any industrial uses of the coastline be "coastal dependent." Storage tanks were questionable in this regard. Moreover, building storage tanks on the pier's fill material seemed to leave them vulnerable to serious seismic hazards. Yet placing them further inland increased air-pollution impacts.
- *Oil Spills.* Frequent oil tanker traffic in and out of San Pedro harbor would greatly increase the possibility of an oil spill.
- *Land-Use Conflicts.* The city of Long Beach and the port were planning to attract more tourist and recreational trade to the area. Approving a major oil terminal with large storage tanks appeared incompatible with these efforts.

The Outcome

A year and a half into the approval process (mid-1977), Pactex had not progressed very far. No major permits had been granted, and SOHIO had only begun to submit completed applications. State and federal agencies worked separately on similar endeavors.

The federal government, concerned that midwestern and eastern energy needs would not be met, became involved in monitoring and coordinating government approvals through the Federal Energy Administration (FEA) and its successor, the Department of Energy (DOE). One positive sign was that the state EIR was nearing completion after being delayed over a dispute regarding the appropriate "lead" agency.[6]

Air-pollution-control permits were most controversial. Confusion reigned over implementing the new offset policy,[7] particularly regarding calculations of emissions, the amount of trade-offs required, and the sources of emissions that were appropriate for trade-offs. SOHIO, without any facilities of its own in the area to clean up, had to negotiate with reluctant third parties for offsets.

Environmentalists and adjacent communities demanded that the company comply fully with applicable regulations and leave the air cleaner than it found it. State air and coastal agencies clashed over siting policy, with the former favoring dispersal of sources, and the state coastal plan requiring concentration in already industrialized areas.

As time passed, a small number of Long Beach citizens became convinced that the project's adverse impacts outweighed its benefits. They formed the Citizens Task Force on SOHIO and began to attack the project from all angles. The task force challenged the state EIR, certified in January 1978. The suit unexpectedly bounced from one court to another over jurisdictional matters relating to the original lead agency dispute.[8]

Opponents also argued that proposed air-pollution offsets were insufficient. Existing air patterns showed that Long Beach would receive most of the pollution and very little benefit from the offsets. Finally, through the initiative process, local opponents had the port's lease with SOHIO placed before the voters in a municipal election. Facing the uncertainty of a referendum over Pactex, SOHIO delayed all regulatory proceedings. A few months later, SOHIO won the election by a large margin through a campaign for "more jobs and cleaner air."[9]

In early 1979, however, company officials began to reevaluate the project. An innovative offset package, including a provision to supply a major utility with a scrubber to clean up sulfur dioxide emissions, had been agreed to and sanctioned by the state air agency, but it faced an uncertain battle before the independent regional air-quality district.[10] The state coastal commission had issued its permit for the overall project early on, without provisions for constructing storage tanks on the pier; SOHIO was seeking an amendment to allow this. Because of unanticipated changes in natural gas supplies, El Paso Natural Gas Company began reconsidering the loss of its 650-mile pipeline.[11] Pending and threatened litigation by local opponents over permit approvals delayed the prospects of construction for an indefinite amount of time. Project costs had doubled to $1 billion, and SOHIO claimed it had spent around $50 million in the permitting process.

SOHIO adopted a two-pronged strategy. First, it pursued a legislative solution, seeking project exemptions from litigation and regulatory requirements at both state and federal levels similar to those obtained from Congress for TAPs.[12] Second, SOHIO told regulators whose agencies had not issued approvals that if decisions were not forthcoming within nine months (by September 1979), the company would terminate the project.

But SOHIO did not get the immediate legislative response it sought from either level of government. Company officials recalculated projections on the North Slope oil; they could not foresee completing construction until 1983, yet North Slope oil production was expected to be at peak levels only through 1985 and then to decrease steadily each year after that. These new calculations lowered profit expectations of Pactex measurably.

SOHIO withdrew the project in March 1979, even though only a handful of permits remained to be granted (and were near approval). Immediately all the parties blamed each other for the project's demise. SOHIO left open the possibility of resuming if approvals could be given within six months. By May 1979, all major permits had been granted, and California passed legislation limiting and expediting judicial review of permit approvals.

Congress was considering additional expediting legislation when, toward the end of May 1979, SOHIO killed the project once and for all, citing increased costs, the prospect of litigation over permits, a declining economic return, and the uncertain availability of El

Paso's gas line, among other reasons. SOHIO's difficulties focused congressional attention on proposals for an energy mobilization board (EMB) to "fast-track" the regulatory process for priority energy projects by imposing time constraints on state and local decisions and waiving certain environmental permit requirements.

Analysis of Regulatory Difficulties

Time took its toll on Pactex. With the flow of TAPs oil guaranteed for only a short time, the project passed the point of profitability from SOHIO's view. Also, SOHIO became more and more frustrated with the regulatory process and less willing to make concessions at the final, crucial stages. No single event or individual defeated it; a variety of factors did.

Although SOHIO's preproject planning was more thorough than Dow's, the company, like Dow, caused many of the project's delays. Pactex's team had little decision-making authority. It was not familiar with the state or local governments. Technical expertise to supply permit information and refute opponents' claims seemed diffused.

SOHIO delayed the regulatory process numerous times, waiting nearly a year before submitting crucial permit applications, neglecting to provide supplemental permit information, and continually making changes in project details that forced new analyses of the environmental impacts each time. Permit procedures were postponed by SOHIO for six months in anticipation of the local referendum campaign and election.

These delaying actions, say some state officials, plus a penchant for seeking quick political solutions to complex regulatory problems and a lack of foresight on air-pollution issues combined to anger regulatory officials and further delay proceedings.

In other areas, SOHIO took risks and suffered unexpected consequences. For example, its attempt to expedite the EIR lawsuit backfired when California courts spent over a year wrestling with the novel procedural issues the company presented. The substantive claims in the case had not even been addressed when the project ended.

SOHIO's actions and delays created doubts about its commitment to build Pactex. Speculation was fueled that SOHIO was more interested in getting Congress to lift the export ban on Alaskan oil to Japan or in gaining congressional exemptions for project permits than in meeting regulatory requirements.[13]

Other factors at state and local government levels collided to produce project delays and weaken prospects for its approval. New agencies and regulatory strategies came into existence just as SOHIO was beginning the new permitting process. A 1976 law changed markedly the California coastal commission's powers and regulatory guidelines, adding new terms and emphases to its authority. It also created new institutional relationships with local governments and permitting procedures. Yet in addition the new law contained special policies on port development that, in fact, made part of SOHIO's task easier.

As SOHIO was preparing its air permits, a new regional air-quality board was established. The law creating it left ill defined the respective authorities of the regional body and the state air-resources board over stationary sources. Adding to this was the promulgation of a new federal rule requiring offsets before a new source of air pollution could be built in an area with dirty air. With little guidance from the federal government, implementation of the offset policy was left to the feuding agencies.

Some officials saw Pactex as a burden on the state and the port area. Press releases issued by state agencies equated Pactex's air emissions to those of a million new cars and declared that the abandonment of the natural gas pipeline would mean a loss of both jobs in southern California and a relatively clean energy source.[14] Thus, these officials focused on getting the most environmental benefits out of Pactex to compensate for serving as an "energy window" for other parts of the country.[15]

Other state actions caused crucial delays. No efficient state mechanism existed to resolve conflicts between the coastal commission and the air-resources board. Further, the Governor's Office of Planning and Research could have prevented the year-long delay in the EIR lawsuit if it had carefully tackled the lead agency controversy and resolved it.

The federal government's lack of leadership compounded delays and frustrations. The Bureau of Land Management prepared the federally required EIS independent of the virtually identical state EIR. The federal EPA remained on the sidelines while the state and regional air agencies grappled with the federally required offset policy.

Also important, while the project was heralded by many federal officials as being in the national interest, supplying urgently needed oil to the Midwest at lower cost and less risk than shipping it through the Panama Canal, the FEA failed to coordinate and oversee siting efforts until two years into the process, and then it only monitored events.

FEA and DOE were decidedly cautious, even timid—sometimes intentionally. Given the political rivalry between Governor Brown and President Carter, some in these federal agencies feared that state officials would oppose the project if federal officials became strong advocates.

Other federal officials claimed that the national interest in the project could not be proved until suitable measures to mitigate adverse environmental effects had been found. These mitigation measures, particularly the air-pollution offsets, took time to negotiate.

Once the state officials were convinced that California's environment could be protected, they were more forceful project advocates. But this came late in the process, roughly six months before the March 1979 withdrawal.

Determined, effective questioning of the project by opponents contributed to an extended permit process. The League of Women Voters challenged the air-pollution impacts of the project, forcing the regional air-quality board to hold lengthy hearings. The Los Angeles City Attorney's Office, on behalf of the neighboring locality, demanded that the offset policy be enforced to result in better regional air quality. The Citizens Task Force on SOHIO, the only group formally opposed to the project, fought Pactex every way possible—on air and coastal permits, the EIR lawsuit, and the local referendum, to name a few.

Significance of Regulatory Delays

Although regulatory difficulties in Pactex's permitting process were important in killing the project, they can be overstated. Considering incomplete applications, postponement of proceedings, and its own changes in the project, SOHIO had given the permitting process roughly two years to approve the project before it pulled out in 1979.

But as SOHIO began to reevaluate the project, it was not the time that had passed that concerned it as much as the time that lay ahead. The regulatory process was winding down, to be completed within a month or two, when Pactex was withdrawn. The litigation process, however, was just beginning. Given pending and threatened suits, the time for project construction became more indefinite, and no one knew how long it would take for cases to make their way through the courts. Thus, SOHIO's decision to abandon Pactex was more than a matter of regulatory delays: litigation posed uncertainty.

Moreover, SOHIO miscalculated the economic time frame of the project. Its profit-making potential was much shorter than previously forecast. Finally, with increasing supplies of natural gas, El Paso Gas Company expressed reluctance to abandon its pipeline. El Paso began negotiating itself out of the loosely constructed agreement with SOHIO, possibly leaving Pactex without a pipeline.

However, SOHIO's siting experience did lead to some positive, if indirect, results, although they are often overlooked amid the uproar that followed Pactex. For example, amendments to the National Environmental Policy Act (NEPA) were passed, allowing federal EIS requirements to be satisfied by similar state laws in some situations.[16]

Once the lead agency dispute over the state EIR was resolved, a task force was established to comment on and supplement portions of the EIR as it was drafted. Composed of all the major permitting agencies, this task force enabled the EIR to be completed within 18 months. The concept has been used successfully since Pactex, and it became one of the models for the recently developed federal "scoping" procedures in NEPA.

Pactex was among the first major projects in California to require the use of third-party air offsets (the new source cleans up an existing source owned by another firm), a policy that had very few implementing guidelines. Pactex negotiations contributed to an understanding of the numerous and complex steps involved in setting up a third-party offset transaction. Successful siting in California and elsewhere has subsequently occurred using offsets.

In retrospect, the project might have demonstrated that large energy facilities can be built without sacrificing environmental quality. Because of the strict enforcement of the offset policy, the air would probably have been cleaner than it was before SOHIO arrived and a new technology (the advanced scrubber), which industry had long resisted, would have been tried.[17] Additional coastal recreational facilities would also have been made available by SOHIO. Moreover, tankers that burned the cleanest fuel would have been utilized—a requirement not in existence for the tankers currently refueling at the port for their journey to the Panama Canal.

SOHIO itself learned a great deal about siting from its experience with Pactex. Recently, it completed the regulatory process for a large petrochemical facility in Texas, meeting all time deadlines. Delays and conflicts were reduced by using high-level management throughout the process. Unlike Pactex, where SOHIO chose the site unilaterally, state officials and environmentalists played an integral role in site selection. After selecting four possible locations, SOHIO let them choose the most acceptable one.

Oil and Oysters—The HRECO Oil Refinery[18]

Dow and Pactex were followed by a headline energy-related siting dispute in the East. Involving the construction of a major refinery along the Virginia coast, this battle received national headlines and is cited as an example of how environmental red tape thwarts the country's drive for energy independence. As of September 1983, the oil-and-oysters controversy continues in the courts.

The Project

The Hampton Roads Energy Company (HRECO) in 1974 announced plans to build a refinery and marine terminal in Portsmouth, Virginia, on the Elizabeth River near the mouth of the Chesapeake Bay. HRECO is 80 percent owned by Cox Enterprises of Atlanta, Georgia, a communications conglomerate, but HRECO's creator is a refinery entrepreneur named John Evans. Evans, in concert with others, has attempted to site refineries in Machiasport, Maine; Savannah, Georgia; and Hawaii.[19] Only the last has succeeded thus far.[20]

The Portsmouth refinery, to be constructed on a 623-acre industrially zoned parcel that HRECO purchased conditionally, would have a refining capacity of 175,000 barrels per day of crude oil and could be expanded to process 250,000 barrels per day. It would produce light products such as gasoline from sour, heavy crudes and would depend on imports from the Middle East. The facility was expected to cost about $350 million by the time it was in operation by 1983.

The Virginia site seemed suited for HRECO's refinery. The company hoped its products would supplant imports to the East Coast from refineries on the Gulf Coast and in the Caribbean and would increase the region's refining self-sufficiency. The site had deep-water access to handle large oil tankers up to 85,000 dead-weight tonnage and was serviced by a good land transportation system. Several alternative sites in the Hampton Roads area were scrutinized by the refinery company before Portsmouth was selected.

In 1974, the city of Portsmouth welcomed the estimated $87 million the project would add to the real estate tax base of $390 million, an addition of over 20 percent. The project, if and when operational, would employ 500 people. Portsmouth's neighbors were less decisive: the cities of Chesapeake and Suffolk supported the refinery, Virginia Beach adopted a neutral position, and Norfolk seemed to oppose it.

The Environmental Regulatory Process

HRECO needed approvals from several federal and state agencies for air-pollution emissions, wastewater discharge, and navigable water dredging. A federal EIS was necessary. The permitting agencies included the Virginia Marine Resources Commission, the state water-control board, the state air-pollution-control board, the federal EPA, and the U.S. Army Corps of Engineers.

Most of the permits were applied for in 1975 and 1976, and the last were not issued until late 1979 and early 1980. Nearly all have been litigated. All except two cases were decided in favor of the refinery, and those two are now (as of September 1983) on appeal. The central issue in this controversy is weighing the need for and the benefits from the refinery against the risk the refinery and related marine traffic pose to the important shellfish and recreation resources of the Chesapeake Bay estuary.

Citizen opposition to the refinery has been led by a Norfolk-based group called CARE, Citizens Against the Refinery's Effects. CARE was organized in January 1976, largely by residents of Norfolk, the city across the river from the refinery site. CARE was the successor group to Tidewater Refineries Opposition Fund (TROF), which had its offices in Newport News, across the Hampton Roads harbor from Portsmouth and Norfolk. TROF, established in mid-1974, shortly after HRECO announced that it wanted to build the refinery at Portsmouth, had approximately 40 to 100 active members.

CARE grew quickly, having organized at the time news spread about Kepone contamination of the James River. The Kepone episode suggested to the group's early members that state regulatory agencies had not been careful enough about allowing discharges to Virginia's waters. Other citizens agreed. By 1981, CARE had a mailing list of 7,000.

CARE's arguments were supplemented by opposition from seafood trade associations, various civic groups, and some local medical societies. The refinery was also opposed by the National Marine Fisheries Service in the U.S. Department of Commerce's National

Oceanic and Atmospheric Administration (NOAA), the U.S. Fish and Wildlife Service within the Department of the Interior, and the EPA.

The primary environmental questions concerned:
- Damage to the region's economically important shellfishery due to oil spills from ship accidents;
- Destruction of fish spawning grounds and nesting areas by oil spills;
- Degradation of air quality in the region; and
- Availability of water supplies, wastewater disposal, and the impact of dredging.

The Chesapeake Bay is the largest estuary in the world, providing over a quarter of the national catch of oysters and over half of the soft-shelled clams. It also produces more blue crabs than all other areas of the nation combined. The Virginia Institute of Marine Sciences (VIMS), a state agency, estimates that 75 percent of the oysters harvested in Virginia depend on seed oysters derived from the James River.

The Elizabeth River, on which the refinery site is located, flows into the James downstream from the oyster seed beds; however, tidal flows could wash oil upstream. The seed beds are regarded by VIMS as unique and irreplaceable, and attempts to replicate them have been unsuccessful. Furthermore, much of the female blue crab population of the bay "overwinters" in lower sections of the bay area nearby or in the channels used by ships serving Hampton Roads.

The Chesapeake Bay is also the primary spawning area on the East Coast for striped bass; 90 percent of the stripers found from Maine to North Carolina spawn there. Moreover, the bay and its wetlands are a major stop along the Atlantic Flyway for migratory birds and waterfowl. Over 500,000 Canada geese winter there each year.

Meanwhile, the Portsmouth area by 1974 was already in violation of federal standards for photochemical oxidants, the primary ingredient in smog (although its air was cleaner than federal standards for several other pollutants). This triggered concern that the air not be further degraded.

People in the area were also concerned about the general impact of the refinery on tourism. The bay is an important recreational resource; the nearby resort community of Virginia Beach alone earns an estimated annual income from recreation and tourism of approximately $92 million.

The Outcome

HRECO secured all necessary local zoning and building permits by fall 1975. Although longer in coming, state environmental permits were also issued for the project, sometimes by closely divided regulatory bodies. The Virginia Marine Resources Commission issued its permit on a four-to-three vote, and the water control board's vote was four to two. VIMS, which advises the Virginia Marine Resources Commission, opposed the refinery, as did the Bureau of Shellfish Sanitation in the Virginia Department of Health, but all state permits were subsequently upheld in court.

Today, HRECO's problems stem largely from the federal permits required. The EPA has approved air- and water-quality permits because the project can operate within all federal statutory and regulatory requirements. Indeed, the emissions from the refinery would be far lower than those from many other refineries, and the wastewater discharges would be cleaner than those required from refineries under prevailing EPA rules. However, in 1978 the EPA, taking into account cumulative environmental considerations and the potential availability of alternative sites with less adverse impact, commented that the U.S. Army Corps of Engineers should deny the project a dredge-and-fill permit. The Corps could approve the project only if it was "in the public interest."

HRECO applied for the dredge-and-fill permit to deepen a channel to its facility in 1975. The Corps, relying on information from HRECO's environmental consultant and from state agencies, published a draft EIS in only eight months. However, the Corps district engineer then denied HRECO's application, citing the threat of an oil spill.

At that point, Mills Godwin, then governor of Virginia, stepped in. He voiced his approval for the project, automatically kicking the decision up to a higher level in the Corps. The Corps district engineer was overruled by the divisional engineer, but objections by the U.S. Fish and Wildlife Service and other federal agencies automatically appealed the decision even higher, to the chief engineer's office in Washington, D.C.

In addition to questions about oil spills, the Corps chief grappled with claims that HRECO had chosen the worst possible site for the refinery. In 1978, at the request of the Corps, an interagency task force examined alternative sites along the eastern seaboard, taking environmental considerations into account. The task force report seemed to support the contention, by environmentalists and others, that the Portsmouth site was the worst possible choice from an environmental perspective.

Corps staff who reviewed the report cited what were said to be gross inconsistencies in it and concluded that the Portsmouth site was acceptable. Consequently, the chief approved the project. The secretary of the Army upheld the decision, over objections by other agencies, on the grounds that estimates of environmental damage were speculative and that there was a national need for additional refinery capacity on the East Coast. Thus, by January 1980 all permits for the project had been granted.

But that was not the end of the matter. The National Wildlife Federation, joined by CARE and other groups, brought suit against the secretary's decision on the ground that it violated the Rivers and Harbors Act, the Clean Water Act, NEPA, and the Administrative Procedures Act. That suit was recently decided in favor of the plaintiffs because of procedural defects in the Corps decision-making process. If upheld on appeal, the decision will probably necessitate the preparation of a supplemental EIS.[21]

A second, and perhaps more important, suit now confronting HRECO was originated by a railroad company. On October 20, 1980, Virginia Holding Corporation, a subsidiary of the Norfolk and Western Railway Company, filed suit against Cox Enterprises. When the railroad sold the refinery site to Cox in October 1974, it was on the condition that construction on the refinery begin within six years of the date of sale. The sales agreement provided that the Norfolk and Western could buy the property back if construction on the refinery did not start by October 17, 1980. Although HRECO cleared some land and began building a fuel-oil storage tank in September 1980, the lawsuit contends that construction did not begin by the deadline and that the Norfolk and Western was entitled to exercise its repurchase option. In August 1981, a Virginia circuit court judge ruled in the railroad's favor; HRECO has appealed.

Analysis of Regulatory Difficulties

The Congressional Research Service (CRS) asked several of the participants in the Portsmouth dispute to identify the bottlenecks they had seen in the permitting process.[22] CRS concluded that the major problems were: failure to identify alternative sites in a timely manner, differences in interpretation of statutory requirements and perception of risk, and, most important, continuing uncertainty about the real long-term need for the refinery.

More specifically, the debate over the risk of oil spills was a crucial one. Although spill prevention and clean-up plans would be prepared, and although many meteorological, ecological, and people-related factors would have to coincide to cause a catastrophic accident, there remained real concern that if an accident occurred the seed beds supporting Virginia's oyster industry and the blue-crab fishery could be wiped out.

Does the Portsmouth controversy suggest that different federal and state administrative procedures should have been used?[23] The interagency federal review ultimately produced good information, but it was too long in coming. Had the "scoping" requirements promulgated by the Council on Environmental Quality (CEQ) in its NEPA EIS regulations of 1978 (explained in chapter 6) been in place when the refinery project started, perhaps more information would have been available earlier in the decision process. Even with scoping, however, interagency agreements would have required a multilevel review process. Nonetheless, the review process might have been expedited if the information file had been more complete earlier.

Conceivably, the state and federal agencies, together with HRECO's consultant, could have agreed in advance that impacts on the shellfishery were a significant issue, and studies of potential impacts might have been undertaken. There also could have been agreement on criteria for evaluating a broad range of alternative sites. Such agreements might have speeded the process of interagency consultation at the federal level and provided a sounder administrative record for the state permitting agencies.

The air-quality permitting process for the refinery left a great deal to be desired. HRECO's proposal was caught up in the changes in federal air-quality programs that occurred during the mid-1970s. Changing guidelines on the prevention of significant deterioration[24] and emerging offset requirements undoubtedly imposed substantial administrative costs on the company. EPA's inexperience with offsets probably contributed to the considerable time required for final federal approval of the state permit for the refinery.

However, air-quality permitting problems alone did not delay this project, because the permit process paralleled the long interagency consultation that preceded final issuance of the Corps of Engineers' permit. Even had no further air-related regulatory action been required after the initial state air-pollution permit was issued in October 1975, work on the refinery would still have awaited resolution of difficult water-related issues in the permitting process of the Corps.

The fragmented character of the state permitting process and the delays in the federal process raise the issue of whether the correct officials were assigned permit responsibilities. The refinery cut across the jurisdiction of many agencies, and no state regulatory officials were authorized to examine all the impacts of the proposed project simultaneously. The state regulatory agencies most concerned with marine impacts were uncertain about the scope of their responsibilities; clearly, they were not encouraged by HRECO to take a broad view.

The Virginia Council on the Environment, which advises the governor on environmental issues, did not have statutory authority to make a balancing judgment, nor did it play a mediating role. In representing the state in federal regulatory proceedings, the council attempted to coordinate the positions of the competing state agencies but expressed neither support for nor opposition to the Corps of Engineers' permit. But in late 1978, the council strayed from its neutral position when, in a letter to the Corps, it supported issuance of the federal dredging permit. As mentioned above, the council's statement of support was made over the continued opposition of VIMS and the Bureau of Shellfish Sanitation, and evidently was based on the council's view that Virginia's governors had previously expressed support for the permit. Thus, state agencies were at odds over whether the project should have been approved.

The final balancing judgment regarding the refinery effectively rested with the federal government. Since administrative delays were most associated with federal permits, the federal agencies opposing the Corps permit became the heavies in this controversy. But the conflicts at the federal level simply mirrored the state-level conflicts.

At the federal level, it was most appropriate for the secretary of the Army, rather than district or division Corps engineers, to make the final decision on the federal dredging permit, since the major justification for the refinery was supraregional—the alleged need for greater refinery capacity on the entire East Coast. The secretarial decision was made after due consultation with other presidential appointees—the secretary of the interior and the administrator of NOAA—having principal statutory authority for protecting marine resources.

It is not clear that mediating efforts by the federal CEQ would have improved the federal decision-making process, except insofar as they might have encouraged early scoping. Section 309 of the Clean Air Act provides for referral to the CEQ of unresolved interagency environmental disputes. (For this particular case, however, involvement by the president's environmental advisory body might have been awkward. The mayor of Portsmouth chaired the Democratic Party in Virginia [and was elected lieutenant governor in 1981]; HRECO's attorney was also well connected politically [also a Democrat, he was elected Virginia's attorney general in 1981]; a major shareholder in Cox Enterprises had been named ambassador to Belgium; and Cox's Atlanta newspapers had strongly supported President Carter's election. These political connections were an undercurrent in public debate over a federal permit for the project.)

Might major institutional or administrative changes have been desirable to facilitate a decision on the HRECO refinery? Even if some consolidated hearing process had been used to evaluate the initial HRECO project proposal, and even if a comprehensive permit for the original proposal had been awarded by a super-siting agency, the proceeding would have had to be reopened, with attendant delay, because HRECO changed the project's wastewater treatment plans.* This change meant that a major institutional reorganization would not necessarily have speeded the state permit process. Furthermore, without some type of prehearing scoping activity and state EIS-type disclosure document, it is doubtful that such a comprehensive hearing would have effectively considered information on oil spills, or that interested citizens would have had a reasonable base of information on which to comment.

Although the federal and state permitting processes were inefficient and sometimes contradictory, HRECO also shares responsibility for the problems the project encountered. The company evidently failed to develop crucial information so that regulators could make a reasonable, timely decision. Earlier development of information combined with better coordination of fragmented state and federal actions would have gone a long way toward speeding a decision.

Additionally, HRECO never made a compelling argument that there

---

* The refinery had first planned to discharge its wastewater to a regional treatment facility. When HRECO decided later to treat its wastewater itself, it was obliged to obtain a state National Pollution Discharge Elimination System permit for its discharge of treated water.

was a national need for the project. Federal agencies—particularly DOE—remained on the sidelines until near the end, exacerbating this failure. Absent such a showing, and in light of the possible damage to resources, regulators balked at approving the project.

Opposition by environmentalists and some local citizens slowed the project, and, in the end, litigation they initiated may actually kill it. But opponents raised many legitimate issues that may not have had a full airing otherwise because those who questioned the project were not included in early project planning.

Even in a less-regulated climate, however, the legitimate concerns revealed in the review process might have surfaced anyway. Moreover, some non-environmental factors, such as market changes and intercorporate struggles, helped cause the demise of not just this project but Dow's plant and Pactex as well.

NEW RULES OF THE GAME

Dow, SOHIO, HRECO—these celebrated siting disputes left corporate headquarters nationwide buzzing. Politicians were stunned. The political and corporate worlds learned the hard way that there are now new rules to the industrial siting game, rules made in response to the effects of indiscriminate plant siting in the past.

Today, companies face the prospect of obtaining scores of environmentally related permits when they decide to build a major new plant or expand an existing one. These laws, regulations, and involved procedures cause a lot of complaining. Some of it comes from people who are fed up with regulations in general. "Congress has just passed too damn many laws," claims Murray Weidenbaum, former chairman of the President's Council of Economic Advisors. "There is a limit to the amount of regulation the American business system can absorb without fundamentally weakening or changing it."[25]

Some of the complaining comes from people who want to go back to the "good old days," but more comes from industry people who accept the need for environmental protection but who are exasperated by a regulatory process that they say can take years and millions of dollars to get through. They cite environmental regulatory "horror" stories as examples of what they are up against in trying to build new plants and expand existing ones that will sustain this country's economic footing.

Ironically, conservationists are not happy either. Many continue to advocate a strengthening of existing laws. At the same time, however, they see indications that the complicated process may actually be hindering the laws' effectiveness in protecting the environment. Environmentalists worry that inefficiency in procedures and permitting may be the opening that their adversaries need to weaken or alter the very substance of those laws. They realize that the United States' recent economic problems make anything that might stand in the way of jobs and capital investment suspect.

The stage has thus been set for a monumental clash between industrial growth and efforts to protect the environment, made more immediate by the fact that the U.S. economy is showing strong signs of recovery from its latest recession.

REFERENCES

1. The three U.S. studies presented in this section are highly summarized versions of much longer accounts prepared for The Conservation Foundation's Industrial Plant Siting Project. The full versions can be obtained from the Conservation Foundation. The case studies are based on numerous interviews, site visits, and in-depth reviews of transcripts, public and corporate documents, and secondary literature. Each has been reviewed for factual accuracy by the people who were active in the disputes.

In addition to examining these three case studies, Conservation Foundation staff tested their conclusions for this report with experts from around the United States who were familiar with the siting process. The researchers also studied hundreds of reports prepared by other organizations and reviewed numerous case studies presented in secondary sources to see if their results agreed with The Conservation Foundation's. The most useful of these reports included:

- Camella Auger and Martin Zeller, *Siting Major Energy Facilities: A Process in Transition* (Boulder, Colo.: The Tosco Foundation, 1979). This major study of energy facility siting prepared by the Tosco Foundation for the Rockefeller Foundation, the Tosco Corporation, and the states of Delaware and New Mexico included a detailed study of siting programs in nine states.
- Urban Systems Research and Engineering, "Causes of Delay for Major Industrial Projects." This unpublished report prepared for the Council on Environmental Quality was based on a study of 41 siting cases.
- John Pinkerton, *Experience of the Forest Products Industry in Obtaining Air Quality Permits for New and Modified Sources* (New York: National Council of the Paper Industry for Air and Stream Improvement, Inc., 1981). This report, based on a survey of 65 projects, is summarized by Pinkerton in an article with the same title in *Journal of the Air Pollution Control Association,* vol. 31, no. 11 (November 1981), p. 1163.
- Doug Linkhart, *Permitting and Siting of Energy Projects: Causes of Delay and State Solutions* (Denver, Colo.: Western Governors' Policy Office, 1981). This report on the siting of energy projects and causes

of delay was based on case studies and papers presented at a December 1980 conference convened by the Western Governor's Policy Office.
- Robert Bander and Abby Pirnie, *Industry Project Planning and the Permitting Process* (Washington, D.C.: U.S. Environmental Protection Agency, 1980). This study, based on a survey of 40 firms, focused on industry project planning and the federal environmental-permitting process.
- David Morell and Grace Singer, eds., *Refining the Waterfront* (Cambridge, Mass.: Oelgeschlager, 1980). This book by two Princeton University professors presents case studies of energy facility siting.
- James Mahoney and Barbara Goldsmith, *Effects of the Clean Air Act on Industrial Planning and Development* (Washington, D.C.: The Business Roundtable and Environmental Research and Technology, Inc., 1981). This book includes 92 case studies prepared by Environmental Research and Technology, Inc., for a Business Roundtable study.
- U.S. Bureau of Land Management, *A Review of the Bureau of Land Management's Energy Facility Permitting Process* (Washington, D.C.: Bureau of Land Management, 1981). This evaluation of the Bureau of Land Management's permitting process, prepared under the direction of Associate Director Dean Bibles, is one of the most thoughtful, balanced intra-agency regulatory-reform studies prepared in recent years.

2. For a more detailed history of the Dow case, see Christopher J. Duerksen, *Dow vs. California: A Turning Point in the Envirobusiness Struggle* (Washington, D.C.: The Conservation Foundation, 1982).

3. Richard A. Liroff, *Air Pollution Offsets: Trading, Selling, and Banking* (Washington, D.C.: The Conservation Foundation, 1980), p. 1:

> The trading of pollution offsets allows industries to site or expand in areas currently violating national ambient air quality standards established by the Clean Air Act. The industries must compensate for the new pollution they will add to an area by either cleaning up emissions from existing polluters (including their own operations) or by buying and shutting down existing polluters, so that the emissions eliminated are greater in quantity than the new emissions added. In other words, the new pollution added is "offset" by the old pollution eliminated. The offset policy allows industrial development while permitting continued progress toward achievement of national standards for ambient air quality.

4. This case study was researched and written by Michael Mantell, an attorney and associate at The Conservation Foundation.

5. Donald Bright, "Sohio Crude Oil Pipeline: A Case History of Conflict," *Transportation Law Journal*, vol. 11 (1980), pp. 243, 255. For a detailed account of the Pactex project, see also a series of reports by Steven R. Thomas prepared for use in Executive Programs in Health Policy and Management, Harvard University, in 1979. These reports include "SOHIO Goes to California," "Tom Quinn and the California Air Resources Board," and "The Standard Oil Company of Ohio Seeks to Build A Facility in California's South Coast Air Basin."

6. A lead agency is the public agency that has principal responsibility for the environmental impact report process. This includes drafting, holding a public hearing, and certifying the EIR as complete. Generally, the lead agency will be the

agency that has primary responsibility for carrying out or approving a project.

7. See footnote 3 for a description of the offset policy.

8. *Citizens Task Force on SOHIO v. Board of Harbor Commissioners*, 591 P.2d 1236 (Cal. 1979).

9. Long Beach voters approved the SOHIO port of Long Beach lease agreement by a 61 to 39 percent margin. SOHIO spent nearly $800,000 in the campaign, compared with $15,000 spent by project opponents.

10. The South Coast Air Quality Management District (SCAQMD) was created in 1977. It has jurisdiction over air quality in southern California, with primary responsibility over stationary sources. The California Air Resources Board (CARB) is the statewide air agency. It has primary responsibility over motor vehicle pollution and oversight authority over all actions taken by the SCAQMD. For example, if the SCAQMD grants a new stationary source permit, the CARB will review it and affirm, deny, or modify SCAQMD's decision. The SCAQMD thus was less independent than the Bay Area Air Pollution Control District that was involved in the Dow case study.

11. SOHIO also faced difficulty securing a pipeline right-of-way on another segment from the Morengo Indians.

12. In order to expedite the building of the trans-Alaskan pipeline, Congress passed special legislation exempting the project from further compliance with the National Environmental Policy Act and severely limiting judicial review of federal government actions taken on the project. 43 U.S.C., sec. 1651-55 (1976).

13. As part of the TAPs project, Congress had prohibited the export of any Alaskan oil, largely in response to concerns that larger tracts of the Alaskan landscape would be put in jeopardy to satisfy foreign, particularly Japanese, energy needs. Others claimed that the oil companies involved in TAPs would find it more profitable to sell Alaskan oil to Japan, a large and more convenient market, thus subjecting the American consumer to potentially higher Middle Eastern oil prices.

14. Bright, "SOHIO Crude Oil Pipeline," pp. 243, 258, 260.

15. Ibid., p. 252.

16. 42 U.S.C., sec. 4332 (D)(1980).

17. The assumption that air might have been cleaner is predicated on the use of a scrubber and other offsets agreed to by the state, not the early proposals advanced by SOHIO.

18. This summary is taken from a more detailed case study researched and written by Richard Liroff, a political scientist and senior associate at The Conservation Foundation.

19. For a description of Evans's activities in Maine, see Peter Bradford, *Fragile Structures* (New York: Harper's Magazine Press, 1975).

20. Evans's Hawaiian refinery was awarded the first annual Blue Sky Award of the American Lung Association of Hawaii for its "extra efforts and accomplishments" in controlling air pollution.

21. *National Wildlife Federation v. Marsh*, Civil No. 80-2350 (D.D.C. July 26, 1983).

22. Congressional Research Service, Library of Congress, "Energy Development Project Delays: Six Case Studies," Senate Committee on Environment and Public Works, 96th Cong., 1st sess., October 1979, Serial no. 96-7.

23. Local governments played only a minor role in the dispute and for that reason are not discussed here.

24. The PSD (prevention of significant deterioration) program established by the Clean Air Act limits the amount of new pollution that can be added by major industrial projects in areas where ambient air quality is superior to national ambient quality standards.

25. Quoted in Craig Garner, "How to Get Better Value for Our Regulatory Dollar," *Context*, vol. 8, no. 3 (1979), p. 6.

Chapter 3

The Cost-Competition Factor

Just how are environmental laws thought to impede industrial growth and new plant construction? One basic way might be called the cost-competition factor. The large amounts of money that domestic firms must spend to comply with environmental and land-use laws allegedly puts them at a disadvantage in competing with foreign firms, drives up inflation, and stifles productivity. Some experts claim that the added cost of complying with regulations has led to a flow of capital investment abroad and discouraged construction of new plants or expansion of existing facilities. Others maintain that enforcement in some states has also led to a competitive shift within the United States that favors the growing Sunbelt over the declining Frostbelt.

The debate over the total cost of environmental regulations, particularly federal pollution controls, is a long-running one. Available data focus on costs imposed by federal pollution-control regulations. In 1978, the President's Council on Environmental Quality (CEQ) estimated that federal environmental regulations would cost industry in excess of $100 billion in investment and operating outlays by 1988.[1] According to a controversial Business Roundtable study, it would cost industry $414.6 billion between 1970 and 1986 to comply with the Clean Air Act alone! While some critics say industry's record of estimating costs seems based on the "Chicken Little" school of economic analysis, even the U.S. Environmental Protection Agency's (EPA's) estimate, which is substantially lower—$291.6 billion—is hardly chicken feed.[2]

Critics maintain that the high cost of meeting environmental-quality regulations puts American companies at a serious disadvantage in competing with foreign firms.* Congress has expressed its anxiety

---

* Critics also argue that environmental-protection costs have either stoked inflation or slowed productivity as industry increasingly put its scarce capital funds into "nonproductive" pollution-control devices. But the debate over the anticipated overall costs of regulations and their impact on inflation and productivity, while insightful, does not offer much practical information about how individual firms actually building new facilities are affected. Moreover, the issues of inflation and productivity have already been studied in some detail by others. Several studies have found that federal pollution-control compliance costs might contribute from

on this point in several ways. One example is in the federal Water Pollution Control Act Amendments of 1972, which directed the Commerce Department to study the competitive impact of environmental regulations on U.S. companies. The amendments also required the president, as a means of heading off any competitive disadvantages, to "undertake to enter into international agreements to apply uniform standards of performance for the control of discharge and emission of pollutants from new sources" through multilateral treaties, the United Nations, and other international forums.[5]

Responding to such fears by copper producers in his congressional district, Representative Morris Udall, a strong advocate of environmental-protection legislation, introduced the Copper Environmental Equalization Act (H.R. 3267, 96th Congress, 1st Session). The legislation would have added about $.10 to the price of copper imports to make up for the rise in U.S. copper production costs allegedly attributable to strong domestic pollution controls. That proposal languished, but it reflects that fear about the impact of such laws on domestic firms.

The specter has also been raised that industries may shop around within this country for states—allegedly those in the Southwest and South—willing to give them an easier time in the environmental and land-use regulatory process. Although officials from Sunbelt states hotly deny these charges, statements by some industry representatives seem to lend credence to them. For example, in the wake of several major environmental permitting disputes a senior vice-president of Dupont told a chemical industry audience that the company "probably will not build in California or any other state with a permit procedure that is as cumbersome. We'll go where we are welcome."[6]

MYTH: ENVIRONMENTAL QUALITY REGULATIONS CAUSE INDUSTRIES TO FLEE TO OTHER COUNTRIES

Are domestic environmental regulations so costly and difficult to comply with that U.S. firms are shopping around for "pollution havens" or shutting down because of an inability to compete with industry from countries with lax laws?[7]

0.2 to 0.6 percent annually to inflation, a small but not insubstantial amount.[3] In another report, a team of University of Wisconsin economists concluded that only 8 to 12 percent of the recent slowdown in productivity growth (which fell from 3 percent annually in 1965 to around 1 percent in 1978 and to a negative figure in 1981) was attributable to environmental regulations.[4]

Since the early 1970s, observers within the advanced industrialized countries have speculated that the costs of meeting environmental regulations will cause a major international redirection in locating pollution-intensive industries. Some academics, policymakers, and even a few environmentalists have said that a significant number of U.S. heavy industrial plants, faced with more exacting and burdensome domestic environmental standards, will seek locations where such regulations are less demanding. The dean of the New York University Graduate School of Business Administration, writing in 1973, was one of the first to raise a red flag. "We would expect environmental pressure to promote a gradual shift of pollution-intensive firms of economic activity from higher income to lower income countries internationally, with the range of activities gradually widening over time."[8] He was careful not to cry crisis and recognized the need for additional analysis. But others picked up his warning and went further, predicting a substantial impact on investment patterns.[9]

Some observers viewed this shift as a positive development. They claimed it would reflect the increasing desire for service-sector and high-technology jobs in the advanced industrial countries and would speed the spread of industrial development to the Third World.[10] Still, adverse effects on the U.S. economy were feared. Some expressed concern that U.S. employment, the balance of trade, and even national security would be threatened if many U.S. industrial facilities transferred production abroad in response to high pollution-control costs.[11]

These debates intensified during the late 1970s and early 1980s as low output and high unemployment continued in many large industries hit with high regulatory costs and as an emphasis on stimulating U.S. exports grew. The finger was pointed at environmental regulations not only for increasing costs to American consumers and depressing the financial outlook of many domestic industries, but also for making it more difficult to compete with firms from other countries.[12]

A 1978 study went so far as to allege that the United States was on the verge of a "wholesale exodus" of major industries, with U.S.-based firms already exporting entire industries—those manufacturing or handling particularly hazardous and toxic substances such as asbestos, arsenic, mercury, and benzidine dyes—to rapidly industrializing countries like Mexico, Brazil, Korea, Ireland, India, and Romania. The report, by a Washington-based chemical engineer and environmental consultant, observed that "the economy of hazard export is emerging as a driving force in new plant investment in many hazardous and polluting industries."[13]

Indeed, during the late 1970s anecdotal evidence suggested that U.S. firms were beginning to search for "pollution havens" outside of the United States and the most advanced countries of Europe.[14] This is illustrated by letters sent in 1978 to operators of American foundries by representatives of several Indian foundries. One letter noted that:

> We understand that because of strict antipollution laws over there in the U.S. many foundries are finding it difficult to cope with setting up of costly equipment for pollution control and as such may be desirous of setting up a part or whole of the equipments in another country. We therefore offer such possibility and invite willing entrepreneurs to Calcutta, where large potentialities for castings in iron exist because of plentiful supply of raw materials and abundant cheap skilled labor.[15]

Letters like these, coupled with attention-grabbing media reports or newspaper articles[16] focusing on alleged examples of flight, refired political concern. Robert Strauss urged representatives at the 1978 Geneva trade talks to consider international environmental and workplace-health standards in light of a "pattern of flight." "American standards in these areas are among the highest in the world," he claimed, "and we do not want this U.S. willingness to protect the environment and our workers to disadvantage the various U.S. producers willing to pay such costs."[17] Labor organizations in the United States also expressed concern that American jobs were being exported to pollution havens.[18]

Exploring the Issues[19]

There is no question that industries spend far more on pollution control in the United States and other advanced industrial countries than in most of the rest of the world. Data collected annually for McGraw-Hill and the Commerce Department make this clear.[20]

It is equally clear, however, that the costs of pollution control are only a small fraction of total capital investment and production costs.[21] This is true even of the capital-intensive manufacturing industries—primary metals, chemicals (including rubber and plastics), pulp and paper, and petroleum refining—that have together accounted for over two-thirds of all capital expenditures for pollution abatement in the United States in recent years.[22] Environmental-control costs for these selected pollution-intensive industries came to 6.2 percent of capital costs for the chemical industry, 6.4 percent for paper, 8.5 percent for petroleum, and 13.5 percent for the primary metals industry in 1981.

Thus, even if environmental-control costs could be completely eliminated in a foreign country, savings would not greatly reduce total capital costs. For less pollution-intensive industries, savings would, on average, be still less. Offsetting these savings would be any extra costs of building and operating overseas. As the Commerce Department has pointed out, total costs for plant construction and operation generally run much higher abroad than in the United States.[23]

Besides, overseas investment decisions are heavily influenced by considerations other than production costs.[24] Access to markets, raw material supplies, political stability, availability of infrastructure and skilled labor, and quality of life considerations for executives are some of the most important of these. Put in context, then, environmental control costs for the most pollution-intensive industries amount to 15 percent or less of capital costs, which are only a portion of total production costs (along with labor, raw materials, transportation, etc.), and total costs are only one of a number of tangible and intangible factors that determine whether a company sets up production facilities abroad.[25]

It should not be surprising, therefore, if many companies choose to stay put. Technical innovation, use of new raw materials or substitute products, reclamation of waste materials, tighter process and quality controls, and other adaptations prove better responses for many companies than flight from regulations. Transportation costs, availability of raw materials, and product standards (which have uniform impacts on all producers for the U.S. market, whether located here or abroad) often favor continued production in the United States even when environment-related problems are formidable.

Still, is there evidence that environmental regulations have influenced the decisions of U.S. industries to build new plants overseas? On the basis of early research for The Conservation Foundation's Industrial Siting Project, we concluded in 1979 that there was not. "Although the volume of foreign investments by American corporate affiliates continued to grow during the 1970s, overall investment patterns do not appear to have been different than they would have been if the United States had not undergone a revolution in its environmental standards."[26]

Several other recent studies reached similar conclusions.[27] For example, one study after reviewing behavioral characteristics of several firms, the factors that generally weigh heavily in foreign direct investment (FDI) decisions, and the current empirical evidence, decided that:

> Flows of foreign direct investment do not appear, as yet, to differ substantially from what would be expected in the absence of environmentally-induced shifts except in a few instances. More importantly, we do not expect a flow of environment-induced FDI of any real significance to materialise in the future. . . . A slight shift at the margin has, indeed, been introduced into the locational calculus of FDI, but for most MNCs [multinational corporations] the shift will not be significant enough to counterbalance the higher costs and risks involved in seeking out developing nations' "pollution haven" for major new facilities. MNCs will, of course, try to locate in areas in which all costs (energy, environmental, political, labour, transport, etc.) are minimised. FDI will, of course, continue to flow to developing nations, especially those with stable governments and indigenous natural resources, but for intrinsic reasons largely unrelated to lower environmental costs.[28]

To test our own conclusions and those of other researchers,* we examined aggregate data on recent overseas investment and trade patterns of the pollution-intensive industries already mentioned—primary metals, chemicals, pulp and paper, and petroleum production. If environmental controls are causing industries to flee, the impact should be most detectable on these polluting industries.

Analysis of that data yields several broad conclusions supporting those of other recent studies.[30]

- During the past decade, U.S. manufacturing industries as a whole expanded their investments abroad as fast as, or faster than, the pollution-prone industries did.
- The percentage of overseas investment going to the so-called less developed countries did not increase to any significant degree for these industries, which have paid the bulk of all pollution-control costs.

---

* A comprehensive study of German companies that have invested in developing countries parallels findings by U.S. researchers. As with American companies, the German firms indicated that the most important factors in their investment decisions were market-oriented rather than related directly to production costs. In addition, many said that labor was the most important cost of production. This study could find no evidence of the existence of pollution havens luring native firms away from Germany.[29]

- The countries favored for overseas investments by the chemicals, metals, paper, and petroleum refining industries were mainly other highly industrialized, heavily regulated countries. It is unlikely that firms "fleeing" U.S. regulations would move to other countries with environmental regulations as, or nearly as, strict.
- Although the total value of U.S. imports grew rapidly during the past decade, import values for the chemicals, metals, and paper industries did not grow significantly faster. They accounted for about the same total share of U.S. imports in 1981 as in 1971. (Refined petroleum imports are obviously difficult to use as an indicator.) Moreover, the percentage of imports of goods produced by these three industries coming from developing countries remained relatively stable during the decade.

Hence, the industrial groupings most likely to be pushed abroad by environmental regulations have not experienced any widespread trends toward relocation in countries with substantially less stringent environmental controls.

Examining individual industries within these broad groupings (for example, polyvinyl chloride within the chemicals grouping, copper within the metals grouping), Conservation Foundation research found that nearly all industries have adapted to environmental regulations with technological innovations, changes in production processes or raw materials, more efficient process controls, and other adaptations that have proved more economical and less drastic than flight abroad. Nevertheless, environmental and workplace-health regulations do appear to have influenced a small number of industries to move more production facilities abroad in recent years. The industries most affected seem to fall into two basic categories.

First, domestic production of a handful of highly toxic, dangerous, or carcinogenic products, such as asbestos, benzidene dyes, and a few pesticides have been disrupted by environmental and worker-protection regulations along with growing public concern and litigation. These industries have not as yet been able by technological adaptation to develop safer substitutes or to meet environmental, workplace-health, and consumer standards easily.[31]

Second, environmental and workplace-health regulations have affected basic mineral-processing industries such as copper, zinc, and lead, as well as a small number of producers of organic chemical intermediates. In these industries, environmental factors have combined with other changing locational incentives (raw material availability, international product cycle) to speed the international dispersion of the industry.

The important point, however, is that healthy, growing American industries are not being forced to move abroad as a result of environmental regulations or public concern in the United States. Of the few industries experiencing international dislocations as a result of these standards, most are also experiencing declines in product demand, lags in technological innovation, and inadequate profits for substantial new capital investments. Thus, isolated examples of industrial flight are not likely to have a significant adverse effect on the U.S. economy in the future; nor are they likely to contribute in any significant way to the industrialization of countries trying to build their industrial base.[32]

Summary

The research for The Conservation Foundation's Industrial Siting Project failed to turn up any credible evidence that environmental regulations have precipitated, or are about to precipitate, a widespread exodus of American industry. In decisions about whether to build abroad or continue operating a facility in the United States, differentials in environmental-control costs are generally outweighed by production and other capital costs. Moreover, other traditional locational factors such as access to markets, proximity of supplies and natural resources, and political stability are almost always far more important than environmental regulations. At most, such regulations affect a decision only when all other factors are equal—which is rarely the case.

Only a handful of U.S. industries (notably, some primary-metal-processing industries and those producing highly toxic chemicals such as asbestos and benzidine dyes) have located more industrial facilities abroad in response to a combination of workplace-health standards, pollution controls, and a variety of nonenvironmental factors. These industries typically are experiencing static or reduced demand as a result of product obsolescence or hazards related to product use.

MYTH: ENVIRONMENTAL LURES LEAD TO INTERSTATE INDUSTRIAL FLIGHT[33]

Another myth associated with the cost-competition issue is that of industrial flight from one region to another of the United States—

that environmental regulations are a significant cause of the Frostbelt-Sunbelt shift.[34]

The issue of regional competition based on weak environmental laws and lax enforcement has nagged at Congress since the 1960s. Congress felt that without uniform federal environmental standards industries could play one state against another by threatening to locate where laws were the weakest— "environmental blackmail," so to speak.[35] The adoption of geographically uniform federal laws quieted the debate over environmental blackmail during the early 1970s, but later in the decade the issue flared again.

Critics—including some politicians from the Northeast and Midwest—claim that some southern states are simply not enforcing federal environmental laws as they should. For example, in 1977 when Dow scrapped its major petrochemical complex in California, citing alleged permitting delays, it was rumored that the governors of several southern states quickly telephoned Dow executives and assured them of smooth sailing through their states' regulatory processes.

Congress has responded to these concerns by amending the Clean Air Act to direct the EPA to determine what regulatory steps are needed to assure regional consistency in standards and procedures.[36] Recently, however, the concerns of Frostbelt politicians have been heightened by Reagan Administration moves to loosen federal oversight of state pollution-control programs.[37]

Even though lax environmental-quality regulations are not an international lure for industry, might they still be used by state governments to attract firms within this country? Have environmental quality laws spurred the Frostbelt-to-Sunbelt movement or contributed significantly to it?

Industry is more "footloose" than ever before. As industrial-location experts point out, classic locational factors such as proximity to final markets or resources are being eroded.

> The industries that have contributed the greatest number of new jobs in recent times have, typically, not been bound to locations near the source of their input of raw materials or even to final markets (e.g., electronics and aircraft manufacturers). Moreover, changing technology has tended to make most industries more footloose. The development of trucking, air passenger, and freight services has permitted the movement of people or goods in a fraction of the time formerly required; miniaturization has reduced bulk; air conditioning has permitted location in a variety of geographical areas previously considered undesirable; and the efficiency and virtual ubiquity of modern telephone communication has further loosened older locational ties.[38]

It is also becoming apparent that, although federal laws are essentially uniform as written, techniques for monitoring and calculating pollution impacts—both critical components in the review of environmental permit applications—are subject to a fair amount of finagling and negotiation. A staff member of the Texas Air Quality Board observed:

> By law there is one EPA. In practice, however, there are some 12 or 13 EPAs [referring to the U.S. EPA's regional offices]. . . . Further, there is a large variance apparently from one state to the next as to the adequacy of state legislation or the degree to which the state agency or governmental entity can conform to the federal act.[39]

Other knowledgeable people agree. A former deputy EPA administrator believes that opportunities for variation "doubtless will also lead in many instances to the government staff making extremely ad hoc judgments on complex technical matters."[40]

Certain key standards and requirements of federal environmental laws, particularly the Clean Air Act, do vary legally by place. For example, plants trying to locate in so-called non-attainment areas—places where air quality violates federal standards—must agree to clean up pollution at existing plants so that the net effect is neutral, a potentially costly requirement not imposed in places where the air is clean. Such a requirement could work against older manufacturing centers with existing air-pollution problems. Likewise, plants locating near national parks and monuments are subject to "nondegradation" requirements of the Clean Air Act, which give those areas special protection.

State and local land-use controls, increasingly at the center of plant-siting disputes,[41] also can vary greatly across the country. Federal incentives have spurred many coastal states to adopt strong laws regulating coastal development. An increasing number of states, including Florida, Wyoming, and Pennsylvania, have direct controls on large industrial and energy projects. In contrast, other areas of the country, particularly rural areas, do not have even rudimentary zoning laws.[42]

Exploring the Issues

Are the rumors of regional competition based in reality? Interviews with industry officials and location specialists found no one willing, even in confidence, to paint one state or region as weak on environmental protection. In fact, several pointed out that projects in states characterized as pro-business have run into serious problems, both with state agencies and grass-roots citizen groups. The proposed BASF chemical plant in South Carolina was killed after a heated battle; oil refineries and chemical plants in Texas have run into regulatory hurdles; and the Hampton Roads Energy Company's major refinery complex in Virginia has had rough going due to local opposition.

Traditional locational theory also seems to argue against a conclusion that environmental laws have a powerful impact on location. As one recent study of most *Fortune* 500 corporations concluded:

> Six considerations are most prevalent in controlling the decision: labor costs, labor unionization, proximity to markets, proximity to supplies/natural resources, proximity to other company plants, and quality of life. The importance of these are often dictated by the nature of competition in the company's industry.[43]

This stance is supported by practically every text written about industrial siting in the past decade.[44] The experts rarely even mention environmental quality regulations in their works.

Although environmental regulations are an important component in business costs, they are apparently not significant enough to outweigh traditional locational factors in most instances. To illustrate, a 1977 survey of the 1,000 largest U.S. industrial corporations by *Fortune* magazine asked key executives to pick the decisive location factors for a specific plant they had sited in the last five years. Only 11 percent ranked environmental considerations ("state and/or local posture on environmental controls and processing of environmental impact reports") among the top three to five factors.[45] Twenty-eight percent mentioned "community receptivity to business and industry," but traditional locational factors—transportation, labor supply, and proximity to customers—were cited far more frequently.

Another study of industrial location decisions by 242 foreign manufacturing investors in the United States asked the respondents to rate the importance of 32 variables in siting new plants. The five most highly rated location factors, in order of importance, were availability of transportation services; labor attitudes; ample space for future expansion; nearness to markets within the United States; and availability of suitable plant sites.[46] Only the final category could conceivably have a relationship to environmental regulations. Interestingly, the respondents ranked attitudes of government officials 18th and, when asked to mention other factors that they thought important, *none* mentioned environmental or land-use controls specifically.

A 1981 study of the location decisions of the 410 largest domestic manufacturing companies, prepared for the U.S. Department of Housing and Urban Development (HUD), identified six controlling concerns in plant location (in no particular order): labor costs, labor unionization, proximity to markets, proximity to supplies/resources, proximity to other company facilities, and quality of life in an area. The study—conducted by interviewing company officials and surveying 562 plant extensions and relocations—concluded that "only if a location passes successfully through one or two of these screens does a company then examine it on other, less constraining grounds."[47]

Environmental permits fell into "factors of second or third level of importance." Seventeen percent of companies opening new plants cited environmental permits as an important regional or state consideration; only 8 percent of relocating firms did the same. When it came to a particular site, only 3 percent of relocating firms mentioned environmental considerations as a factor. Of firms closing plants, 8 percent mentioned compliance with environmental and workplace-health regulations, which ranked a poor ninth behind inefficient plant layout or outmoded technology, high wage rates, and price competition, among other factors, and just above problems with raw material prices, supply, and transportation.

The preliminary results of a project at the University of Cincinnati—though slightly different from the *Fortune*, foreign investors, and HUD findings—provide the most up-to-date support for the conclusion that environmental laws are not significant lures.[48] After analyzing how frequently certain terms or phrases were used by respondents, the researchers have drawn tentative conclusions that are illuminating:

- Environmental considerations are more important locational influences than they were in 1970, but they have by no means supplanted the major traditional location factors.
- Geographic variations in the cost of pollution-control equipment are relatively unimportant. Variations in uncertainty about obtaining permits, the time required, and the number of permits are more crucial.
- Companies do not explicitly search for areas with lower environmental-control costs, although they do appear to be very attracted by states that offer efficient permitting procedures. Ease in permitting was linked closely in respondents' minds to

a state's business-climate image.
- Environmental regulations are far more important in selecting a site within a region than in selecting a particular region in which to locate.
- Decision-makers do not believe they are operating much differently than they did a decade ago with regard to the number of alternative sites seriously considered in the location search.

How states present themselves to new industry also reveals the relative importance of environmental controls in plant location. Three-fourths of the states, as well as many substate regions, have some type of industrial development agency. These agencies advertise the virtues of their areas in business periodicals. If jurisdictions believed that differences in environmental controls or in amount of procedural red tape gave them a locational advantage, they would publicize it in some (perhaps veiled) way.

A survey of advertisements by state development agencies found that only 1 of 25 advertising states alluded to the environmental regulatory process.[49] That instance was Georgia's reference to "our one-stop environmental permitting service." In only one other instance (a brief mention by Massachusetts that pollution-control bonds were available) was pollution even mentioned. On the other hand, West Virginia and New Mexico made explicit reference to the quality of life and recreational opportunities that relocated employees might find there.[50]

Five states did give prominence to their "good business climate," and five others mentioned it in passing.[51] South Carolina argued that "We could paint a picture of South Carolina by painting just one word . . . profit," while Ohio maintained that "In Ohio we believe profit is *not* a dirty word." Despite this emphasis, none of the states, save Georgia, made any linkage between business climate and the permitting process.

Another indication that states do not try to compete on regulatory issues is found in a survey of several thousand industrial development organizations. Respondents were asked to rank eight "positive factors" for economic development in their area, such as transportation and labor availability. In three successive years (1976-78) "lack of red tape in obtaining environmental permits" was rated last (least positive) in all but three states.[52]

Are these survey results, which generally rank environmental concerns fairly low on the locational-factor scale, borne out by experience? Is there any evidence of a movement of pollution-intensive firms from states with strong controls to those with a more lenient attitude? One early study, based on data gathered from 1968 through 1972, combined pollution-control costs in each state with other locational factors in a regression equation constructed to explain locational shift. Pollution-control factors were calculated to have a marginal but significant effect on locational shifts of heavily polluting industries.[53] But that study was hindered by data limitations. Also, it focused on a period before uniform federal pollution-control laws had time to make their impact felt. No current analysis is available.

The commonsense approach to assessing the impact of environmental-control variations on industrial location is to look at the actual locational choices of large, highly polluting industries over the past decade. However, that approach has at least two limitations. First, although environmental regulations have been quite prominent since about 1970, some of those thought most likely to influence industrial siting are of a more recent vintage. Second, reliable data on new plant locations, broken down by state and region, are not publicly available.

California, for example, did not implement its stringent coastal-zone land-use controls until 1973. The EPA did not begin requiring plants locating in dirty-air areas to seek compensating pollution offsets until late 1976.[54] Other potentially important regulations, aimed at preventing significant deterioration of existing air quality (so-called PSD regulations), were only getting off the ground in the late 1970s.

Although it is possible to look at the variation over time in employment by industry and by location, it is impossible to separate out how much of a given change was caused by new plants, closure of old plants, or expansion or contraction of the labor force of existing plants. Thus, analysis of location data must make the rather risky assumption that each of these phenomena is affected in the same way by environmental regulations.

Despite these limitations, an attempt was made in 1979 to look at the statistical record.[55] Data from published federal sources (County Business Patterns) on the location, by state, of industrial employment during the years 1970 and 1976 were used because the period between those two years was one of dramatic extension of federal and state pollution regulation, as well as an active time in the field of land-use control. These data were used to calculate, for each state and for each industry considered, the "competitive shift" in employment between 1970 and 1976. (The competitive shift is the difference between the actual change in employment in a state and the change that would have occurred had industry grown or declined in that state at the average national rate.)[56] The working hypothesis of the investigation was that states with relatively lax environmental land-use policies would have garnered a higher-than-average competitive shift in industrial employment, particularly in pollution-intensive industries.

To obtain a measure of regulatory severity/laxity, a rough index was constructed. The scores of 48 states (Alaska and Hawaii were excluded) on this index ranged from 2 to 17, with a high value indicating the most stringent environmental regulatory climate. The rankings were consistent with intuitive feelings about various states' regulatory severity. For example, California, Maine, and Minnesota ranked as the three strictest regulators, with high values on the index, and Alabama, Louisiana, and Mississippi ranked as the least severe.

Table 3.1 shows how the competitive shift relates to this environmental regulation (ER) index. Considering manufacturing as a whole, states having a positive competitive shift in the 1970-1976 period had a mean ER index value of 8.7. States showing a loss had a mean value of 9.9. Thus, the "lax" states had a slight (but not statistically significant) tendency to gain jobs.

Paradoxically, for pollution-intensive industries, the difference was even less pronounced. After those industries were separated, in this analysis, into five sectors—paper, chemicals, petroleum refining, rubber and plastics, and metals processing—gainers had a mean ER index value of 8.8; losers had a value of 9.4. The greatest differences in employment between gainers and losers occurred in primary metals and in chemicals. In both cases, the states gaining employment had, as measured in this index, the laxer regulations. For metals, the mean difference between states was statistically significant.

This work was updated in late 1982, focusing on the period 1976 through 1980. The results were comparable, as shown in table 3.2. The states that gained employment in pollution-intensive sectors had slightly lower ER scores than those that lost employment. This difference was about the same for both periods examined.

Next, the results for 1970 to 1976 were compared with those for 1976 to 1980. During the later period, the gainers had noticeably higher ER scores than did the losers (9.5 versus 8.1), even though the difference was still not statistically significant. However, in the aggregate 1976-80 rankings for pollution-intensive industries, the dif-

Table 3.1
Shifts in Industrial Employment, 1970-76
(by state, excluding Alaska and Hawaii)

MANUFACTURING AS A WHOLE

Greatest Gainers	Net Job Gain	ER Index*	Greatest Losers	Net Job Loss	ER Index
Texas	131,493	9	New York	300,358	11
California	110,454	17	Pennsylvania	140,780	10
North Carolina	67,017	7	New Jersey	85,922	11
Mississippi	47,340	3	Ohio	75,554	11
South Carolina	47,177	10	Illinois	71,556	6
34-gainer average		8.7	14-loser average		9.9

FIVE POLLUTION-INTENSIVE INDUSTRIES

Greatest Gainers	Net Job Gain	ER Index	Greatest Losers	Net Job Loss	ER Index
Texas	22,477	9	New York	34,108	11
North Carolina	17,529	7	New Jersey	25,350	11
South Carolina	16,889	10	Pennsylvania	24,342	10
California	16,084	17	Ohio	18,229	11
Arkansas	10,276	7	Illinois	16,410	6
30-gainer average		8.8	18-loser average		9.4

PAPER

Greatest Gainers	Net Job Gain	ER Index	Greatest Losers	Net Job Loss	ER Index
Maine	9,541	14	New York	9,758	11
Arkansas	3,366	7	Pennsylvania	3,378	10
California	3,164	17	Massachusetts	4,265	13
Tennessee	2,637	6	Ohio	3,970	11
Alabama	2,602	2	Michigan	3,720	7
31-gainer average		8.7	17-loser average		9.5

CHEMICALS

Greatest Gainers	Net Job Gain	ER Index	Greatest Losers	Net Job Loss	ER Index
North Carolina	9,017	7	New Jersey	11,658	11
South Carolina	7,755	10	New York	7,689	11
California	7,228	17	Virginia	5,801	6
Louisiana	4,682	3	Pennsylvania	5,542	10
Missouri	3,461	4	West Virginia	4,352	9
24-gainer average		8.1	24-loser average		9.9

PETROLEUM REFINING

Greatest Gainers	Net Job Gain	ER Index	Greatest Losers	Net Job Loss	ER Index
Texas	5,880	9	Indiana	2,821	10
Ohio	606	11	Pennsylvania	2,527	10
Washington	485	13	California	1,517	17
Alabama	484	2	Kentucky	916	11
Colorado	475	10	Oklahoma	877	4
19-gainer average		9.0	29-loser average		9.0

RUBBER AND PLASTICS

Greatest Gainers	Net Job Gain	ER Index	Greatest Losers	Net Job Loss	ER Index
Texas	7,494	9	Ohio	13,554	11
North Carolina	6,328	7	Massachusetts	11,674	13
California	5,634	17	New York	7,996	11
South Carolina	4,550	10	Connecticut	7,153	12
Arkansas	4,268	7	New Jersey	4,882	11
29-gainer average		8.7	19-loser average		9.6

METALS PROCESSING

Greatest Gainers	Net Job Gain	ER Index	Greatest Losers	Net Job Loss	ER Index
Kentucky	4,475	11	Maryland	8,804	10
Georgia	4,255	7	New York	7,969	11
West Virginia	3,848	9	Pennsylvania	7,447	10
Missouri	3,581	4	New Jersey	7,403	11
Texas	3,137	9	Illinois	6,882	6
31-gainer average		8.3	17-loser average		10.4

*The ER (Environmental Regulation) Index measures the stringency of each state's environmental regulations. Scores range from 17 (most stringent) to 2 (most lax).

Table 3.2
Shifts in Industrial Employment, 1976–80
(by state, excluding Alaska and Hawaii)

MANUFACTURING AS A WHOLE

Greatest Gainers	Net Job Gain	ER Index*	Greatest Losers	Net Job Loss	ER Index
California	256,384	17	New York	143,162	11
Texas	101,175	9	Pennsylvania	126,809	10
Florida	74,870	10	Ohio	119,143	11
Washington	46,008	13	Illinois	94,175	6
Arizona	43,674	7	Michigan	84,562	7
30-gainer average		9.5	18-loser average		8.1

FIVE POLLUTION-INTENSIVE INDUSTRIES

Greatest Gainers	Net Job Gain	ER Index	Greatest Losers	Net Job Loss	ER Index
Texas	20,172	9	Ohio	39,872	11
California	17,141	17	Pennsylvania	24,742	10
South Carolina	8,093	10	Michigan	14,043	7
Florida	7,228	10	Maine	9,687	14
North Carolina	6,166	7	New York	6,547	11
30-gainer average		8.8	18-loser average		9.3

PAPER

Greatest Gainers	Net Job Gain	ER Index	Greatest Losers	Net Job Loss	ER Index
North Carolina	2,543	7	Maine	10,051	14
Georgia	1,830	7	Louisiana	2,986	3
Texas	1,748	9	Michigan	2,010	7
Kansas	1,247	7	Kentucky	1,544	11
Oklahoma	885	4	Minnesota	1,493	14
28-gainer average		8.9	20-loser average		9.2

CHEMICALS

Greatest Gainers	Net Job Gain	ER Index	Greatest Losers	Net Job Loss	ER Index
Texas	9,119	9	New York	5,417	11
Louisiana	3,816	3	Pennsylvania	4,644	10
California	3,443	17	New Jersey	3,606	11
Kansas	1,590	7	Michigan	3,113	7
Washington	1,460	13	Ohio	2,529	11
29-gainer average		9.2	19-loser average		8.8

PETROLEUM REFINING

Greatest Gainers	Net Job Gain	ER Index	Greatest Losers	Net Job Loss	ER Index
Washington	881	13	Texas	3,050	9
New Jersey	637	11	Ohio	1,319	11
Oklahoma	614	4	Pennsylvania	890	10
Louisiana	608	3	Colorado	524	10
Florida	407	10	Maryland	472	10
35-gainer average		8.7	13-loser average		9.9

RUBBER AND PLASTICS

Greatest Gainers	Net Job Gain	ER Index	Greatest Losers	Net Job Loss	ER Index
California	7,185	17	Ohio	19,676	11
Florida	6,296	10	Indiana	6,472	10
South Carolina	6,035	10	Massachusetts	3,390	13
Texas	5,515	9	Connecticut	2,201	12
Minnesota	3,176	14	Arkansas	1,824	7
29-gainer average		9.1	19-loser average		8.8

METALS PROCESSING

Greatest Gainers	Net Job Gain	ER Index	Greatest Losers	Net Job Loss	ER Index
Texas	6,840	9	Pennsylvania	17,357	10
California	5,737	17	Ohio	15,761	11
Indiana	3,576	10	Michigan	8,328	7
Arizona	3,489	7	Illinois	3,537	6
Washington	2,252	13	Alabama	3,112	2
37-gainer average		9.1	11-loser average		8.5

*The ER (Environmental Regulation) Index measures the stringency of each state's environmental regulations. Scores range from 17 (most stringent) to 2 (most lax).

ferences between gainers and losers (8.8 versus 9.3 ER scores) remained almost unchanged.

This finding can be interpreted as a rather weak indication that during the 1976-to-1980 period the states with more-stringent regulations improved their performances at creating or attracting new manufacturing jobs but did not improve their ability to attract pollution-intensive sectors. Or, alternatively, these findings can suggest that less-stringent regulators for some reason (perhaps growing unionization) lost some of their ability to attract manufacturing jobs generally but kept their ability to attract the polluters.

Applying the same procedure to other data produced similar results. For example, states in which foreign firms made the most large new plant investments between 1975 and 1977 (table 3.3) as a whole had a slightly higher ER index than did those states that received no foreign-owned plants. Looking at individual states, it appears that foreign companies prefer to locate plants in traditionally highly industrialized states, despite their somewhat stricter regulatory climates. (The analysis of foreign plant location was updated in 1982, drawing on the most recent data available from the U.S. Department of Commerce, which covered the years 1978 to 1979. No variation from the earlier findings was found.)

There is also a wide range of values on the 1979 and 1982 ER indexes among both gainers and losers. For example, states making net gains in employment in pollution-intensive industries included such "strict" states as California, Maine, and Oregon, as well as such "lax" states as North Carolina and Arkansas. Similarly, losers included a wide range of both strict and lax states. The same variation was true for states receiving or failing to receive new foreign plants.

This statistical evidence indicates that laxity alone is not sufficient to attract polluters to states.[57]* Nor does a harsh regulatory environment appear to stop firms from locating in jurisdictions where the pull of traditional locational forces has not yet weakened.

---

* Of course, this historical evidence cannot document the future effect of environmental laws that are still in the process of being implemented (particularly, the PSD and offset regulations under the Clean Air Act) or of new laws that may be enacted (perhaps controls over emissions linked with acid rain). Some industry officials claim that they will avoid areas in which offsets are required (that is, areas where air is dirtier than national standards allow) to avoid burdensome monitoring and permitting requirements. In contrast, others say they will shun PSD areas

Table 3.3
New Major Plant Investments by Foreign Companies
(by state, excluding Alaska and Hawaii)

States Receiving Most Plants

New York
California
New Jersey
Texas
Pennsylvania
Ohio
Virginia
North Carolina
Georgia
Michigan
Louisiana

Average ER index = 9.0

States Receiving No Plants
Mississippi
Arkansas
Oklahoma
South Dakota
North Dakota
Idaho
New Mexico
Wisconsin
Nevada
Arizona
Montana
Wyoming

Average ER index = 8.2

Source: Foreign investment data supplied by Ingo Walter.

Other recent studies of industrial location buttress these conclusions. For example, in 1979, the Illinois Institute of Natural Resources, a state agency, published a report analyzing the influence of environmental regulations on industrial plant migrations.[59] Illinois, like many other Frostbelt states, had experienced a serious decline in its manufacturing base during the 1970s, losing 180,000 manufacturing jobs from 1970 to 1979. This decline, which caused concern throughout the state, was attributed by several observers to a bad business climate brought on, in part, by environmental laws that were more stringent than regulations in some other, particularly midwestern, states. However, the Institute of Natural Resources' study challenged that claim, concluding that the state's environmental regulations accounted for only one plant closing between 1971 and 1977.

Another research project tracked 5.6 million firms over a seven-year period to determine, among other things, if the decline of manufacturing jobs in the Northeast and Midwest was caused by firms migrating to the South.[60] Its conclusions: plant closures in the Frostbelt were not linked to new plant openings in the South. Few plants migrated from North to South. The reality was that firms springing up in the South were completely new ones or branches of established companies, not old ones fleeing overregulation and other perceived impediments associated with the Frostbelt.

The previously mentioned HUD study also lends support. One of its main conclusions was that relocations are just a small part of the geographical shifts in manufacturing employment. Only 13.7 percent of all plant relocations by major firms, involving 14.4 percent of relocated employment, crossed state boundaries.[61] This same study found that environmental permits were a factor in very few relocation decisions.

(locations with air cleaner than required by national standards) because of the possibility of time-consuming reviews.[58]

In addition, previously mentioned proposals by the Reagan Administration to devolve more environmental-regulatory authority to the states, and the as-yet-unknown fallout from the Administration's cuts in the EPA's budget, might reopen opportunities for "flexibility" in state administration of federal regulations. That possibility might also affect industrial decision-making in the future.

Nevertheless, given the evidence that environmental regulations are only secondary factors in plant-location decisions, one must remain skeptical that these considerations will have a significant impact on industrial siting.

Summary

The right to pollute is not an important locational determinant. No evidence of a migration of industry from one state to another in search of "pollution havens" was unearthed, although there was a very slight tendency for states with lax environmental and land-use regulations to make relative gains in employment in pollution-intensive industries.

These states made even clearer gains in manufacturing as a whole, including industries unlikely to be affected by environmental regulations in making location decisions, which suggests that states with lax regulatory environments may also have other advantages (low labor costs, Sunbelt location, low unionization) that attract pollution-intensive industries. In fact, the most important aspect of environmental regulation may be its effect on perceptions about business climate. A state's reputation for a burdensome environmental permitting system helps shape a company's notion of the state's overall business climate—often an important factor in location decisions. By the same token, however, a reputation for strong environmental protection may increase a state's attractiveness to dynamic industrial sectors that attach great importance to quality of life.

REFERENCES

1. Council on Environmental Quality, *Environmental Quality—1978* (Washington, D.C.: U.S. Government Printing Office, 1978), p. 429.

2. For an excellent discussion of these and other environmental regulatory cost studies, see Paul Portney, "The Macroeconomic Impacts of Federal Environmental Regulation," in Henry Peskin, Paul Portney, and Allen V. Kneese, *Environmental Regulation and the U.S. Economy* (Baltimore, Md.: Johns Hopkins University Press, 1981). See also Lawrence Mosher, "The Clean Air That You're Breathing May Cost Hundreds of Billions of Dollars," *National Journal* (October 10, 1981), p. 1816; Robert Healy, "Environmental Regulations and the Location of Industry in the U.S.: A Search for Evidence" (Washington, D.C.: The Conservation Foundation, 1979), p.2.

3. These studies are analyzed in Portney, "Macroeconomic Impacts."

4. Gregory B. Christiansen, Frank Golley, and Robert H. Haveman, *Environmental and Health/Safety Regulations, Productivity Growth, and Economic Performance: An Assessment* (Washington, D.C.: U.S. Government Printing Office, 1980). Their work is summarized in Peskin, Portney, and Kneese, *Environmental Regulation and the U.S. Economy*, p. 55.

5. P.L. 92-500 (October 18, 1972), 33 U.S.C., sec. 1251 *et seq.*
6. Statement of Richard E. Herbert to the Manufacturing Chemists Association, Pebble Beach, California, March 1978, quoted in *Chemical Week*, vol. 122, no. 13 (March 27, 1978), p. 34.
7. The discussion that follows summarizes findings presented in three other publications from the project that delve into these questions in much greater detail. See H. Jeffrey Leonard and Christopher J. Duerksen, "Environmental Regulations and the Location of Industry: An International Perspective," *Columbia Journal of World Business*, vol. 15, no. 2 (Summer 1980), pp. 52-69; H. Jeffrey Leonard, "Striking a Balance between a Clean Environment and the Cost of Continued Economic Growth," *Business Facilities*, vol. 16, no. 4 (May 1983), pp. 22-23; and H. Jeffrey Leonard, *Are Environmental Regulations Driving U.S. Industry Overseas?* (Washington, D.C.: The Conservation Foundation, forthcoming).
8. Ingo Walter, "Environmental Management and the International Economic Order," in C. Fred Bergsten, ed., *The Future of the International Economic Order: An Agenda for Research* (Lexington, Mass.: D.C. Heath, 1973), pp. 313-14.
9. Ralph C. d'Arge and Allen V. Kneese, "Environmental Quality and International Trade," *International Organization*, vol. 26, no. 2 (Spring 1972); Eugene V. Coan, Judy N. Hillis, and Michael McCloskey, "Strategies for International Environmental Action: The Case for an Environmentally Oriented Foreign Policy," *Natural Resources Journal*, vol. 14, no. 1 (January 1974), p. 94; Testimony of Ralph Nader before U.S. Senate, Committee on Public Works, Subcommittee on Air and Water Pollution, *Economic Dislocations Resulting from Environmental Controls*, Hearings, 92nd Cong., 1st sess., 1971; Ingo Walter, "Pollution and Protection: U.S. Environmental Controls as Competitive Distortions," *Weltwirtschaftliches Archiv*, vol. 110, no. 4 (1974); and Allen V. Kneese, "Environmental Pollution: Economics and the Policy," *American Economic Review*, vol. 61 (1971).
10. Ralph C. d'Arge, "Trade, Environmental Controls and the Developing Economic," in *Problems of Environmental Economics* (Paris: Organization for Economic Cooperation and Development, 1971); Horst Siebert, "Environmental Protection and International Specialization," *Weltwirtschaftliches Archiv*, vol. 110, no. 4 (1974); and Horst Siebert, "Trade and Environment," in Herbert Giersch, ed., *The International Division of Labour: Problems and Perspectives* (Tubingen: International Symposium, 1974).
11. These fears are summarized in C. Fred Bergsten, "The Threat from the Third World," in Richard N. Cooper, ed., *A Reordered World: Emerging International Economic Problems* (Washington, D.C.: Potomac Associates, 1973), p. 114; and Stephen P. Magee and William F. Ford, "Environmental Pollution, Terms of Trade and Balance of Payments of the United States," *Kyklos*, vol. 30 (1972).
12. Louis J. Cordia, "Environmental Protection Agency," in Charles L. Heatherly, ed., *Mandate for Leadership: Policy Management in a Conservative Administration* (Washington, D.C.: Heritage Foundation, 1981).
13. Barry I. Castleman, "The Export of Hazardous Factories to Developing Nations" (Independent Report, March 7, 1978), p. 3. The report is summarized in *Congressional Record* (June 18, 1978), pp. 3559-67.
14."Exporting Pollution: What Does It Cost," *Not Man Apart* (February 1976), p. 3; "Hazardous Industries Flee to Developing Countries," *Not Man Apart*

(February 1976); Helen Dewar, "Study Cites Firms' Flight to Third World to Avoid Safeguards," *Washington Post*, June 30, 1978, p. C-1; Rep. David R. Obey, "Export of Hazardous Industries," *Congressional Record* (June 18, 1978), pp. 3559-67; and Elizabeth Jager, "Multinationals and Jobs," address to a seminar on Multinationals, Jobs and the Environment, Washington, D.C. (October 20, 1978).

15. These letters, on file at The Conservation Foundation, were provided by the Cast Iron Metals Federation.

16. Bob Wyrick, "Hazard Export" (special ten-part series), *Newsday*, December 16-31, 1981; David C. Williams, "Hazardous Jobs Have Become One of America's Major Exports," *Los Angeles Times*, September 23, 1979; P. Sweeney, "Juarez Plant a Runaway Firm?" *El Paso Times*, April 4, 1978; Edward Flattau, "U.S. Firms Seek Refuge from Regulations," *Chicago Tribune*, June 16, 1979.

17. Quoted in *Environmental Reporter*, vol. 9, no. 10 (July 7, 1978), p. 451.

18. Sheldon W. Samuels, "National Stewardship: Unilateral International Regulation of Occupational and Environmental Hazards: The Position of the Industrial Union Department, AFL-CIO" (Washington, D.C.: September 29, 1980), pp. 62-63; Herman Rebhan, "Labor Battles Hazard Export," *Multinational Monitor*, vol. 1, no. 2 (March 1980); and Barry Bluestone, Bennett Harrison, and Larry Baker, *Corporate Flight: The Causes and Consequences of Economic Dislocation* (Washington, D.C.: The National Center for Policy Alternatives, 1981).

19. A variety of sources have been examined for this analysis. The project team studied macroeconomic investment trends in general and for the pollution-intensive domestic industries that have been hardest hit by control costs. The team also surveyed other current studies of industry investment decisions overseas and conducted a series of field interviews and case studies both in the United States and in foreign countries that have been singled out as possible pollution havens because of their weaker regulatory requirements, proximity to the large markets of the United States and Europe, and apparent success in attracting multinational firms. Finally, the International Institute for Environment and Society in Berlin conducted a parallel study on the location decisions of German industrial firms.

20. For a discussion of the data on the comparative costs of pollution control, see Charles Pearson and Anthony Pryor, *Environment: North and South* (New York: Wiley-Interscience, 1978), p. 170.

21. This issue was first raised in Robert Healy, "Environmental Regulation and Industrial Siting in the U.S.: A Search for Evidence," in *A Conference on the Role of Environmental Land-Use Regulation in Industrial Siting* (Washington, D.C.: The Conservation Foundation, 1979).

22. Gary L. Rutledge and Betsy D. O'Conner, "Plant and Equipment Expenditures by Business for Pollution Abatement, 1973-1980 and Planned 1981," *Survey of Current Business*, vol. 61, no. 6 (June 1981), pp. 19-25.

23. This point is made forcefully with numerous comparative-cost illustrations, in U.S. Department of Commerce, Industry and Trade Administration, *U.S. Industrial Outlook 1980* (Washington, D.C.: U.S. Government Printing Office, 1981), pp. 135-36.

24. See Healy, "Environmental Regulation and Industrial Siting in the U.S."; and Leonard and Duerksen, "Environmental Regulations and the Location of Industry."

25. For a discussion of traditional location theory, see David M. Smith, *Industrial Location* (New York: John Wiley & Sons, 1981) and Howard Stafford, *Principles of Industrial Facility Location* (Atlanta: Conway, 1980). See also Thomas N. Gladwin and Ian H. Giddy, "A Survey of Foreign Direct Investment Theory," Working Paper No. 86 (Ann Arbor, Mich.: Graduate School of Business Administration, University of Michigan, November 1973).

For an attempt to place environmental-control costs in the context of foreign direct investment theory, see Thomas N. Gladwin and John G. Welles, "Environmental Policy and Multinational Corporate Strategy," in Ingo Walter, ed., *Studies in International Environment Economics* (New York: Wiley-Interscience, 1976), pp. 177-224.

The Conference Board has published several studies that explore some of the reasons for recent foreign investments made by American multinationals. See James R. Basche Jr., "Production Cost Trend and Outlook: A Study of International Business Experience," *Report No. 724* (New York: The Conference Board, 1977); and James Greene and David Bauer, *Report No. 656* (New York: The Conference Board, 1975).

For an elaboration of factors other than production cost that influence foreign investment decisions, see James R. Basche Jr., "Foreign Production Costs: A Survey of Recent Trends and Their Effects on Business Policy," Paper No. 697 (New York: The Conference Board, 1976). In fact, it appears that multinational companies have little in-depth knowledge about foreign environmental regulations or potential pollution-control costs at the time that they make an initial decision to locate outside their home base. In many cases, environmental costs and analyses become an influential factor only after a corporation gets to the stage of specific plant design. See Thomas N. Gladwin, *Environment, Planning and the Multinational Corporation* (Greenwich, Conn.: JAI Press, 1977).

Investment guides and services that provide companies with investment reports also offer *prima facie* evidence that multinational corporations are not very concerned with national environmental regulatory requirements in foreign investment decision-making. Surveys of foreign investment rules published by the Economist Intelligence Service, the regional journals of Business International, the American Industrial Properties Report, and other business sources neither evaluate environmental regulatory climates nor provide information on relative environmental-control costs. It is probably reasonable to assume that if this information were in high demand by corporations, the investment surveys would devote more attention to it.

26. Leonard and Duerksen, "Environmental Regulations and the Location of Industry."

27. See Gladwin and Welles, "Environmental Policy and Multinational Corporate Strategy"; Pearson and Pryor, *Environment: North and South*; "Corporate Investment and Production Decisions: Does Environmental Legislation Play a Role," *Economist Intelligence Report* (November 1978); and "Environmental Factors Seen Rarely Decisive in Site Selection," *International Environment Reporter*, vol. 1, no. 5 (May 19, 1978), p. 142.

28. Gladwin and Welles, "Environmental Policy and Multinational Corporate Strategy," p. 202.

29. Gabriele Knödgen, *Environment and Industrial Siting: Results of an Empirical Survey of Investment by West German Industry in Developing Countries* (Berlin: International Institute for Environment and Society, May 1979). Knödgen's findings are presented in *Zeitschrift Für* Unweltpolitik, vol. 2, no. 4 (1979), and in *IfoSchaelldienst*, nos. 1-2 (1981). Knödgen makes several important points in her study:

(a) Both large and small firms have different investment motives. The bigger firms would like to open new markets, or at least maintain their market share, whereas the smaller firms generally would like to reduce their production costs. However, although smaller firms would like to reduce their costs more frequently than larger firms, environmental-cost reduction does not play as important a role for them.

(b) Firms have developed new and more efficient technologies for their production in industrialized countries, and firms install them in developing countries because they expect developing countries to impose stricter regulations in the future. They see a narrowing of any cost gap in the future and want to avoid the high cost of retrofitting plants with modern pollution control technology.

(c) Firms that rank environmental motives high as reason for relocating all belong to basic industries (for example, chemicals and nonferrous metals). They do not, however, cite any preference for one country or region based on environmentally induced investments. Knödgen could find no evidence of the existence of pollution havens.

(d) Firms sometimes use the threat of relocation to developing countries in negotiations with German agencies even if that alternative is unrealistic.

30. The project team decided to investigate whether aggregate investment and import patterns for the pollution-intensive industries were changing in directions that one would expect if environmental concerns had gained a more predominant role in international industrial locational considerations. The team expected to see two changes in foreign investment and import trends: more-rapid expansion of overseas investments and imports for the pollution-intensive industries than for manufacturing industries overall, and higher percentages of the overseas investments and imports of such industries to involve underindustrialized nations with low levels of environmental regulations—that is, the grouping of so-called less developed countries. Although the existence of such trends would not prove that environmental factors have been responsible, their absence, it seemed, would provide strong evidence that environmental factors have not gained the prominence that many analysts predicted. The data to support these findings will be presented in the forthcoming companion report to this volume: Leonard, *Are Environmental Regulations Driving U.S. Industry Overseas?*.

31. Some examples are detailed in H. Jeffrey Leonard, "Environmental Regulations, Multinational Corporations and Industrial Development in the 1980s," *Habitat International*, vol. 6, no. 3 (1982), p. 9.

32. None of these findings means that some U.S. firms, intent on finding a particular foreign location for a new production facility, have not bargained strenuously to reduce their required anti-pollution measures; nor does it imply that U.S. firms operating in other countries have never created pollution problems. Some of the problems and misunderstandings that have resulted are noted in H. Jeffrey

Leonard, "Pollution Plagues Industrial Firms in Growing Nations," *Conservation Foundation Letter* (August 1982). See also H. Jeffrey Leonard and David Morell, "Emergence of Environmental Concern in Developing Countries: A Political Perspective," *Stanford Journal of International Law*, vol. 17, no. 2 (Summer 1981), pp. 281-313. These issues will be dealt with in depth in a forthcoming Conservation Foundation Issue Report on multinational corporations and pollution in rapidly industrializing nations.

33. This section is based on and summarizes conclusions drawn from Healy, "Environmental Regulation and Industrial Siting in the U.S." Statistics and data analysis were updated in 1982.

34. That the Sunbelt states are attracting manufacturing jobs at a furious rate while the Frostbelt states are stagnating or declining is demonstrated in U.S. Department of Commerce, Bureau of the Census, *Statistical Abstract of the United States 1981* (Washington, D.C.: U.S. Government Printing Office, 1982), pp. 780-83.

35. A. Myrick Freeman, "Air and Water Pollution Policy," in Paul Portney, ed., *Current Issues in U.S. Environmental Policy* (Baltimore, Md.: Johns Hopkins University Press, 1978).

36. Interview with Paul DeFalco, Jr., former U.S. EPA regional administrator, Region IX, in San Francisco, August 1979. The regulations on uniformity and regional consistency are required by Section 301(a)(2) of the Clean Air Act, 42 U.S.C., sec. 7402(a). Uniformity regulations can be found at 40 CFR Part 56.

37. For example, the Reagan Administration has proposed that states be able to classify rivers under the Clean Water Act as to use and degree of pollution allowable. This would destroy the current uniform application of national water effluent standards, and, according to environmentalists, "balkanize" the existing regulatory scheme, putting it back to where it was in the 1960s.

38. Barbara Davis et al., *The Effects of Environmental Amenities on Patterns of Economic Development* (Washington, D.C.: Urban Institute, 1980), p. 87.

39. Roger Wallis, unpublished speech to a National Air Quality Conference, San Francisco, January 15, 1979.

40. John Quarles, *Federal Regulation of New Industrial Plants* (Washington, D.C.: published by the author, 1979), p. 38. See also, "Plant-by-Plant Negotiations of Permits Belies Image of Inflexible Regulations," *Environmental Reporter* (September 10, 1978).

41. Nelson Rosenbaum, *Land Use and the Legislatures* (Washington, D.C.: Urban Institute, 1976). Rosenbaum notes that of 12 proposed East Coast oil refineries killed between 1970 and 1974, 9 owed their demise to local objections, not failure to meet pollution-control standards.

42. National Association of Counties and International City Management Association, *The County Yearbook* (Washington, D.C.: NACO/ICMA, 1977) and unpublished tables supplied by NACO.

43. Roger W. Schmenner, *Making Business Location Decisions* (Englewood Cliffs, N.J.: Prentice-Hall, 1982). The subsequent cites to Schmenner's work are from a summary given to this author by the project sponsor, the U.S. Department of Housing and Urban Development.

44. See Smith, *Industrial Location*; Stafford, *Principles of Industrial Facility Location*; Barry Moriarty, *Industrial Location and Community Development* (Chapel Hill, N.C.: University of North Carolina Press, 1980); and Leonard F.

Wheat, *Regional Growth and Industrial Location* (Lexington, Mass.: Lexington Books, 1973).

45. *Facility Location Decisions*, Fortune Magazine Market Research Survey (New York: Fortune, 1977).

46. Hsin-Min Tong, *Plant Location Decisions of Foreign Manufacturing Investors* (Ann Arbor, Mich.: UMI, 1979), p. 57. Similar findings were made in a study that focused on state business incentives: 73.3 percent of manufacturing firms responding to mail survey said air quality regulations were an "insignificant influence." Only 1.1 percent cited them as a deciding positive influence. Michael Kieschnick, *Taxes and Growth* (Washington, D.C.: Council of State Planning Agencies, 1981).

47. Schmenner, *Making Business Location Decisions*, pp. 9, 13-17.

48. Howard Stafford, with Ethel A. Galzerano and James Kelley, "The Effects of Environmental Regulations on Industrial Location: A Summary" (Cincinnati, Ohio: University of Cincinnati, Department of Geography, June 1983). Stafford directs this research. He and several colleagues have conducted an extensive mail survey and have interviewed scores of business executives involved in the recent domestic location of new manufacturing plants. All the individuals were with large, multiplant corporations. The first parts of the interviews were open-ended, allowing respondents to describe freely the decision process and the location factors involved in siting a specific new plant. The interviewees were not told that the research was focusing on the effects of environmental regulations. The second portions were devoted to the respondents answering questions explicitly related to specific variables, including 13 involving environmental considerations. The interviews were taped, transcribed, and subjected to content analysis, a technique that analyzes responses by measuring the frequency with which certain terms or phrases are repeated by the person being interviewed.

49. The advertisements appeared in *American Industrial Properties Report, Office/Industrial Site Seekers' Directory, 1979* (Red Bank, N.J.: American Industrial Properties Report, 1979).

50. Quality of life/recreation was also a frequent theme of advertisements placed by cities and counties.

51. Among them were Florida, Georgia, Ohio, Oklahoma, and South Carolina (major reference), and Connecticut, Nevada, Massachusetts, Vermont, and West Virginia (passing reference).

52. The only states not ranking this factor last (eighth) were Arizona (sixth), Massachusetts (fifth), and New Hampshire (fifth). *Site Selection Handbook*, vol. 23, no. 2 (May 1979), pp. 81-82.

53. Keith Leitner and M. Jarvin Emerson, "Effects of State Pollution Control Programs on Industrial Location," discussed in Davis et al., *The Effects of Environmental Amenities*, pp. 173-74.

54. See footnote 3 in chapter 2 for an explanation of offsets.

55. Conservation Foundation economist Robert Healy undertook this task. The subsequent discussion and data presentation are the results of his work.

The index by Healy was composed of a weighted sum of the following measures: (a) per capita state spending on air-pollution regulatory program administration; (b) the voting record on environmental issues of the state's Congressional delegation—a proxy for statewide voter attitudes; (c) existence of a state power-plant-siting law; (d) existence of a state requirement that local zoning laws be consistent with comprehensive plans; (e) existence of a state environmental-impact-statement process not limited to government projects; (f) the state planning director's ranking of the priority given "environmental protection" by his or her state legislature. Data for the last measure are taken from Richard A. Mann and Mike Miles, "State Land Use Planning: The Current Status and Demographic Rationale," *Journal of the American Planning Association* (January 1979).

Healy's index was expanded in 1983 by project staff to cover 23 indicators. The resulting state-by-state rankings based on this expanded index were very similar to those utilized by Healy and do not change his analysis. The revised index and state rankings are set forth in appendix A.

56. The "competitive shift" and the technique of "shift-share analysis" is explained in Harvey S. Perloff et al., *Regions, Resources, and Economic Growth* (Baltimore, Md.: Johns Hopkins University Press, 1960).

57. Healy points out that these conclusions should be treated cautiously and suggests the use of more-sophisticated statistical models to study the issue further.

58. For a discussion of possible locational impacts of the Clean Air Act, see Arthur D. Little, Inc., *The Impact of National Energy and Environmental Policies on Industry Location Dynamics: A Methodological Review* (Washington, D.C.: U.S. Department of Energy, 1979).

In a twist on the interstate "flight" claims examined by the project staff, Professor B. Peter Parshigian of the University of Chicago asserted (in a 1982 study for the Center for the Study of American Businesses at Washington University, St. Louis) that passage of laws like the federal Clean Air Act resulted in no small part from the desire of northern industrial regions to stop the movement of industry to the South and West. Parshigian claims that the Clean Air Act, pushed mainly by congressmen from the already industrialized North, penalizes plants moving to rural areas because of PSD regulations.

Although PSD regulations do entail complex reviews, Parshigian apparently overlooks the impact of the so-called "offset" rules promulgated under the same Clean Air Act, which require companies locating in areas with dirty air (most industrialized northern cities) to seek compensating offsets from other sources to cancel out any new pollution. This process is certainly as detailed and lengthy as PSD review. Moreover, Parshigian ignores the fact that industries are subject to a wide range of state and local environmental and land-use laws that vary tremendously from jurisdiction to jurisdiction without any clear geographic pattern. For example, California and Florida communities have very strong land-use laws as do those in Vermont, Hawaii, and many areas in Maryland. Parshigian's study is sumarized in the *Los Angeles Times*, November 18, 1982, p. I-C-6.

59. Anthony C. Petto and Frances E. Oxley, *Environmental Regulations and Other Factors Influencing Industrial Plant Locations* (Springfield, Va.: National Technical Information Service, 1979), pp. xi, 2-3.

60. David Birch, *The Job Generation Process* (Cambridge, Mass: MIT Press, 1979). Birch's conclusions are summarized in Neal Peirce, "Smokestack Chasers Who Miss the Point," *Washington Post*, May 26, 1979.

61. Schmenner, *Making Business Location Decisions*, p. 24.

Chapter 4

The Permitting-Maze Syndrome

The debate over the total cost of federal environmental regulations may deflect scrutiny of an even more serious situation: a growing panoply, at *all* levels of government, of environmental-quality laws—sometimes overlapping, sometimes contradictory—that can entangle a project.

While Congress was responding to public concern about pollution in the 1970s, state and local governments were also active. A number of states now have centralized land-use control laws (once the exclusive province of local governments) aimed at protecting "critical environmental areas"—wetlands, floodplains, prime agricultural lands.[1] The states have also imitated the federal government by passing environmental policy acts—little NEPAs (National Environmental Policy Acts)—requiring the filing of independent environmental impact statements (EISs), not only for state projects but also for large private developments in many cases.[2] A majority of states directly control large industrial and utility development.[3]

Local governments have also changed their regulations, both substantively and procedurally. While some areas of the country are still without even rudimentary zoning, others have legislated sophisticated land-use controls on all forms of development.[4] From industry's point of view, the most significant development is not the strictness of such laws but the growing discretion they place in the hands of local officials. To obtain discretionary approvals, industry must often placate local citizens—by agreeing to additional pollution controls not required by federal law, for example, or by contributing to community improvements such as sewers, roads, or other facilities.

The plethora of new laws is more than just a minor annoyance to industry. When businesspeople—those building new plants, mines, and energy facilities—express concerns about siting, foremost on their minds is what they see as the sheer complexity of the environmental-review and permitting system. They say this complexity causes long delays in constructing new facilities, not only adding to project expenses but also making long-term planning for production and marketing uncertain.

In summing up a session on regulatory reform and the Clean Air Act at a major meeting of business executives, one participant said:

> It came out loud and clear that the problem here is not substance but process. As one gentleman representing a major company said, he had four permits awaiting action for seven months. It was the same set of permits with no changes in the technology and in the amount of pollution the new plant would create from their previous permits. And they had been waiting seven months and heard nothing.[5]

The former head of the Department of the Environment in the United Kingdom, who spearheaded an effort to reduce project delays in that country, captured the perplexity of industry: "It has always struck me as curious that a judge can sentence a man to life imprisonment minutes after a jury verdict, yet a planning inspector can take an age to reach a decision after an inquiry."[6]

Some industrial executives have had their pictures snapped while unfurling long charts that detail the environmental-review process their companies must contend with in building a new production facility. The terms *thicket*, *maze*, and *labyrinth* are often found in the accompanying caption. For example, the Construction Industry Legislative Organization of Hawaii studied the delays that development projects were running into in that state. The organization's executive director said:

> We finally tried to illustrate our findings in a flow chart which would show the process from start to finish for a typical project. We knew it was bad, but even we were not prepared for what resulted. When we unrolled our flow chart at a legislative hearing, our lawmakers were flabbergasted. It was 40 feet long. It graphically showed that there was so much overlap and duplication and shuffling of papers back and forth that it was a miracle any project could survive the ordeal.[7]

Delay in permit reviews may hobble industry in several ways.[8] Delay means additional time to complete a project, and time means money. If a company must spend months or years negotiating the so-called permitting maze—getting federal pollution permits, meeting state environmental regulations, complying with local land-use laws—the cost of the project goes up, dramatically so if double-digit inflation exists, as it did in the late 1970s. Although the cost of delay is hard to pin down, several reports argue the effect is substantial.[9]

Delay in permitting, some argue, has another, more subtle and pernicious, effect. Industry representatives claim they are no longer sure that necessary environmental or land-use permits will be issued within a certain period or, worse, that the answers they do receive will stick. Consequently, project planners are less able to establish reliable horizons to satisfy a particular market or to meet future competition.

In some cases, businesses may decide not to invest, given such uncertainty, causing what might be called "stillborn" projects. One survey, the National Council of the Paper Industry for Air and Stream Improvement (NCASI) study of forest-product industry respondents, reported that 16 projects were scrapped in the planning stages because of difficulties feared with the Clean Air Act.[10] As one observer put it,

> Regulations are expensive, but the real drag on the economy is caused not so much by regulations as [by] a regulatory system filled with uncertainties, contradictions, and unnecessary stringencies. It simply has become impossible for businessmen to predict what environmental standards they will be forced to live with, and so plans to build new plants have been delayed or scrapped.[11]

Even though this chilling effect of environmental regulations remains unmeasureable, two other widespread impressions of regulatory impact do appear to be misunderstandings: (1) that environmental and land-use regulations are blocking or delaying so many projects that industrial development is being thwarted, and (2) that other industrial nations get more plants built because their environmental regulatory systems are less onerous and less complex.

MYTH: RED TAPE IS STRANGLING INDUSTRIAL DEVELOPMENT

Hampton Roads, Colstrip, Kaiparowits, Coal Creek, the trans-Alaskan pipeline—these are hardly household names, but they are all familiar, legendary horror stories to people in industry charged with building new plants. From a sampling of headlines, articles, and speeches, one might think that industrial projects no longer get built in the United States, thanks to environmental laws and other assorted regulatory ills.

One study, based on a comprehensive survey of newspaper and magazine articles, claims that between 1970 and 1978 half of all proposals for new or expanded oil refineries were blocked from locating at the site originally chosen and 58 percent of petrochemical projects were delayed due to environmental problems.[12] Another reported that delays in issuing 55 dredge-and-fill permits for power plants increased costs by almost $20 million.[13]

A study done in 1981 for the Business Roundtable, a prestigious association of business executives, focused on the impact of the Clean Air Act on industrial planning and development.[14] The report, based on a survey of 92 cases cited by 150 companies as illustrative of regulatory problems associated with the act, studied both procedural and substantive problems. It concluded that the Clean Air Act "imposed unnecessary costs" and "created many avoidable constraints" to industrial development. Further, it alleged that these 92 cases were only "the tip of the iceberg" when it came to regulatory problems plaguing industry.

Exploring the Issues

Is the sheer complexity of environmental review and permitting thwarting industrial development and making long-term planning impossible?

Horror stories about delays caused by environmental regulations obscure the fact that thousands of plants are, in fact, being built. Many of the available studies also seek out problem cases rather than success stories. For the record, it is well to recall that the number of manufacturing establishments in the United States rose from about 311,000 in 1967 to about 360,000 in 1977 (the latest year for which figures are available), an increase of 16 percent.[15]

Most manufacturing plants do not need hundreds of environmental permits, nor are they subjected to the lengthy and complex reviews that, say, a petrochemical refinery is. For example, in 1980 U.S. semiconductor firms' capital spending exceeded $1.5 billion. Although these types of plants and facilities are not without environmental problems,[16] most were built quickly, without serious delay.

The majority of siting problems in the 1970s were experienced by a few heavy manufacturing sectors such as oil refining, petrochemicals, and primary metals processing. Yet each of the most pollution-intensive sectors—pulp and paper, chemicals, oil refining, rubber and plastics, and primary metals—increased its number of establishments between 1967 and 1972 and again between 1972 and 1977, the same period when so many new environmental laws were being implemented (and, again, the latest years for which figures are available).[17]

Total capital investment in heavy manufacturing increased from about $35 billion in 1973 to almost $50 billion in 1980 (measured in 1972 dollars).[18] Undoubtedly a good share of these funds went into existing plants and additions, which generally engender less environmental opposition, but the fact remains that since 1970 industry has continued to build, expand, and modernize.

Over the last decade many industrial projects—including many heavy industrial projects—proceeded with few, if any, regulatory hitches. An evaluation of air-pollution permit applications between 1977 and 1979 by coal-fired power plants, which typically take longer to get permits than most other facilities, reveals that only 2—the Colstrip and Intermountain power-related projects—of 76 applications suffered significant delays. Seventy-four were approved with an average processing time of five-and-a-half months.[19] (These power plants have many of the same problems pollution-intensive manufacturing plants have.)

An industry study surveyed delays in obtaining dredge-and-fill permits from the Corps of Engineers for oil- and gas-related development in southern Louisiana from September 1, 1978, through April 30, 1979.[20] Companies were asked to provide information about the time it took to obtain permits, the costs incurred, and any production lost or deferred. Most objections to the Corps issuing such permits came from the U.S. Fish and Wildlife Service, the National Marine Fisheries Service, and the Louisiana Shellfish Producers Association.

This is the study that, as already noted, said that the reporting companies had experienced almost $20 million in cost increases for 55 permits. The study also shows, however, that 17 of the 40 companies responding reported no permit difficulties. At the end of the study period, only 3 of 76 permit applications had not been granted; the rest were approved and issued on average in about four months.

The federal EPA's 1982 report on air-quality permitting for new facilities is also enlightening.[21] The report focused on permits issued under the prevention of significant deterioration (PSD) program. New facilities being sited in areas with air quality better than national standards for specific pollutants (for example, particulate matter, sulfur oxides, nitrous oxides) must undergo a special review to ensure that air quality in the area is not lowered significantly. The PSD review is widely considered to be one of the most (if not *the* most) complex and time-consuming of all federal environmental regulatory processes. Moreover, during the period covered by the study (1978-1980), the PSD program was in flux due to changes in regulations and to court actions.

Despite these factors, the average processing time for PSD permits was 7.1 months once an application was deemed complete by the EPA, and almost 90 percent were processed within one year.[22] If time is calculated from when applications were first submitted (many were sent back by the EPA because they contained inadequate information or were poorly prepared), the average time is 11.0 months, with 68 percent processed within one year and 88 percent within 18 months.[23] Of course, other federal permitting procedures and those at state and local levels proceed concurrently with PSD review.

As mentioned above, the NCASI surveyed forest-product manufacturers' experiences in obtaining air-quality permits for new and expanding facilities.[24] NCASI's thoughtful, balanced study did iden tify a handful of projects that appeared to be mired in the permitting process—5 projects took over two years each to obtain approval. It also found, however, that 25 of 65 projects covered by the survey had no significant problems. Overall, the median time from air-permit application to approval was three months if a PSD evaluation was not involved and nine months if one was.[25]

It does take longer to build plants today than in the 1960s, and people are quick to blame regulators for the extra project time. In fact, however, gross figures do not tell much. When building schedules are stretched out, the fault is often not solely—or even primarily—that of the regulatory process. Even in the headline-making disputes, when plants are blocked or years pass before ground is broken, delays caused by environmental quality regulations are often less significant than those attributable to financing problems, lack of consensus regarding need, labor disputes, construction and equipment delivery snafus, and regulatory hurdles not associated with environmental protection.

For example, AMAX, Inc., which spent three years planning a massive molybdenum mine in Gunnison County, Colorado, appeared to be close to final regulatory approval when it abruptly announced that the project would be delayed because the molybdenum market had declined dramatically. Similarly, U.S. Steel whizzed through the early stages of planning for a giant new steel facility at Conneaut, Ohio; it killed the project because of other problems facing the steel industry, including overcapacity and competition from cheap imports.

The experience of one utility in securing air-pollution permits further illustrates this point.[26] The utility wanted to switch a plant from natural gas fuel to coal and needed permits from both the U.S. EPA and a state environmental control agency to do so. The entire switchover process took over two years from the date of the permit application until coal was used in the boiler. Obtaining the necessary permits was not easy, and mistakes made by the utility's engineer in calculating emissions exacerbated matters, forcing the company to apply for a permit change days after one was issued. However, the real hang-up was the late delivery of pollution-control equipment (a particulate baghouse), not any difficulties in the permitting process.

Some industry studies, notably the Business Roundtable report already mentioned, have failed to note the full range of causes of delay. In several cases cited by the report, environmental regulations did cause delay, but additional investigation shows that problems caused by poor project planning and other inadequacies attributable to industry were also significant. For example, in some instances companies prepared their permit applications inadequately, causing government officials to balk at approving a project until they had more information. Moreover, the study did not consider the fact that time spent in the permitting process is not always time lost overall. Once permit applications are made, companies typically continue working on other aspects of project planning such as financing, equipment design, and construction contracts.

Another study, undertaken for the Council on Environmental Quality (CEQ) by Urban Systems Research and Engineering, found a mixture of factors contributing to drawn-out project schedules.[27] Even in these cases, which represent some of the most bitter industrial-development/environmental clashes of the 1970s, economic factors played an important role.

For example, while environmentally related problems surfaced in 28 of the 41 cases reviewed, economic factors slowed 25 of the 32 mineral extraction and energy projects and 4 of the 9 nonenergy industrial developments. These economic stumbling blocks included inadequate financing, changing market demand, availability of fuel and labor, and questions about antitrust liability. Declining demand was particularly disruptive for power-generation projects—5 of 11 were so delayed. The study concluded that economic factors were less significant as causes of delay in non-energy industrial projects, although availability of capital was an important impediment. Delays in equipment delivery and labor stoppages were mentioned as problems in a handful of cases, although none was canceled for these reasons.

Other studies that did not focus exclusively on well-known environmental disputes generally concluded that project delays were just as likely to be brought on by factors unrelated to environmental or land-use regulations. A review of Federal Power Commission (FPC) data on delays in power-plant licensing and construction between 1967 and 1976 found, for example, that the principal causes of delay in bringing new power plants on-line were related to labor problems, falling demand, and the inability of suppliers to deliver high-quality equipment on schedule.[28]

Changes in plans and financial problems were found to be more significant since 1974, when, contrary to earlier expectations, demand for power leveled off. Regulatory problems ran a poor third to vendor- and labor-related difficulties, according to the FPC data. Regulatory problems did become more significant in 1973 and 1974, probably because utilities were grappling for the first time with new clean air regulations and restrictions on natural gas and oil usage. Since then, regulatory problems have declined in importance.

Two industry-funded reviews lend credence to the CEQ and FPC studies. A 1979 review of overall reliability and adequacy of North American power systems by the National Electric Reliability Council cited both regulatory and financial uncertainties as the biggest utility challenges in the 1980s.[29] The National Coal Association's 1978 survey of future electric-utility capacity additions concluded that "a prime reason for any delay in 'on-line' dates is, of course, a downward revision in the utility's load growth forecast."[30] The report cited regulatory and financial uncertainties as major secondary reasons.

Even the most ardent environmentalists must recognize that regulations do sometimes delay new plant construction. There is no persuasive evidence, however, that the problem is as severe as some in industry have alleged. Studies that seek out the worst cases, cannot, without additional investigation, properly conclude that these cases are the "tip of the iceberg," indicative of a widespread problem,[31] as the Business Roundtable study does.*

---

* The Business Roundtable study, as already noted, examined 92 cases in which industrial development was affected by air-pollution regulations. While this study was better documented than many recent industry-financed studies,[32] its research design elicited problem cases from industry. As previously explained, delays in some of the 92 cases cannot be blamed entirely on regulators; industry must share the responsibility. Moreover, 92 cases represent a tiny fraction of the total number of industrial projects subject to environmental and land-use reviews each year. Although the federal EPA does not compile statistics on the total number of air-

Furthermore, delay caused by the regulatory process may be essential to protect legitimate public interests. Complainers often use the term *delay* as a loaded word that connotes "unnecessary." They seem to assume that time needed to evaluate project impacts can somehow be magically reduced to zero. In case studies written by industry organizations, *delay* is often used to characterize time devoted to essential preconstruction environmental monitoring or future impact assessment. Before deciding that delay is in fact unnecessary, it is important to pose several questions: Was the time devoted to project evaluation used productively? Did the evaluation benefit society by avoiding or mitigating threats to prime wildlife habitats, for example, or risks from air pollution? Did the industry benefit by making adjustments to the project (perhaps locating an alternative site) that reduced costs and opposition in the long run? Could the same results have been attained more quickly and efficiently?

Finally, in many instances in which environmental or land-use regulations delayed a project, problems can be attributed to what might be called "teething pains." Both industry planners and government regulators were, during the 1970s, faced with complying with or implementing novel and complex laws, and this very newness caused some delay. For example, Dow's proposed chemical plant in California was a guinea pig for two new laws: a new air-pollution policy that would allow facilities to be built in dirty-air regions, and California's agricultural-lands protection act. What was learned in the Dow case, both by industry and government regulators, was invaluable in later, successful siting attempts by other companies across the nation.

Similarly, after SOHIO canceled its proposed California-to-Texas oil terminal/pipeline project, it put the environmental planning lessons it learned to work to swiftly site a petrochemical complex in Texas. Instead of choosing a site on its own, the company invited governmental officials and citizen groups to help select a location that would be acceptable to them. That approach helped defuse opposition.

Overall, there is evidence that the number and intensity of environmental conflicts in the United States are slowly decreasing as the players in the siting game learn the new rules.

---

pollution permits reviewed, it did grant more than 1,100 New Source Review permits (mostly PSD or nonattainment permits, widely cited by industry as the most difficult air-pollution permits to obtain) between 1978 and 1980,[33] the period the Business Roundtable study examined.

88 PART II: PROBING THE HEADLINE CONCERNS

Summary

A significant number of industrial facilities have been built relatively quickly over the past decade with few or no serious environmental problems. These success stories are often overlooked in the clamor over celebrated siting battles. Even in the headline-making disputes, delays caused by environmental quality regulations are often less significant than those attributable to financing problems, labor disputes, construction and equipment delivery snafus, lack of consensus regarding need, and regulatory hurdles not associated with environmental protection. When regulations do cause delay, that delay may be essential to protect legitimate public interests. Moreover, a good deal of the regulatory delay of the 1970s can be attributed to "teething pains" that are likely to be eased as the players in the siting game learn the new rules.

MYTH: OTHER COUNTRIES ARE BETTER AT RECONCILING ENVIRONMENTAL REGULATION WITH INDUSTRIAL DEVELOPMENT

There are some flaws in the U.S. environmental and land-use regulatory system. But are these flaws unique to the way this country approaches regulation? Some industry officials argue that our Western allies—West Germany, France, Japan, and Canada—are more successful, for various reasons, at getting new projects built. Some say environmental laws abroad are not as strict. Others maintain that governments and industries in Europe have eschewed the adversarial relationship so prevalent in the United States for a more cooperative approach. Is there anything to be learned from the foreign experience?

The myth about foreign superiority in the environmental regulatory game has two sources. First, there is a common belief that several major industrial facilities, blocked in the United States, were later sited in other developed countries. For example, the ill-fated Dow chemical plant in California reportedly was, according to company officials, built later in British Columbia without any major delays.[35] From there it now serves the western United States petrochemical market. The Dow case is often cited by business executives as a prime example of how our industrialized allies do it better.

A second source of the superiority myth is comparisons between specific U.S. and European environmental regulatory procedures. Industry representatives point out that they do not have to worry about overly complex U.S. laws relating to air-pollution control—such as PSD and offset regulations—when building plants in Europe. The result, they claim, is that facilities can be built much more quickly and with greater certainty.

Can these claims withstand scrutiny? Case studies of siting disputes in Europe, presented in the following section, indicate that our industrial allies have some unique problems of their own. Although the United States does have stronger environmental regulation in some areas, notably air and water pollution, some countries, including the United Kingdom, West Germany, and France, have far more restrictive and geographically widespread land-use laws that virtually dictate where a plant can or cannot be built. These case studies also reveal that European nations are tightening many of their environmental regulations, patterning them after U.S. laws.

Like Dow and SOHIO, these European studies are worst-case examples and should not be mistaken for the norm. Still, they show clearly that other industrialized nations have their own peculiar, but significant, problems regulating the environmental impacts of major industrial facilities.

Exploring the Issues: Foreign Case Studies[36]

Barlöcher Chemical[37]

Barlöcher CWM (BCWM), a major West German chemical producer founded in the early 1800s, began searching for a site for a new production facility in the early 1970s when the area around its Munich facility became urbanized, hindering expansion. Although relatively small in size compared to giant firms like DuPont and Dow, BCWM is a dynamic, innovative, and highly specialized chemical firm boasting some 400 patents and exports to 35 countries. Its specialty is heavy metal (for example, lead and cadmium) plastic stabilizers.

The new production facility planned by BCWM would produce various lead compounds, including lead stearate and lead oxide. All totaled, the project, to be built in phases, would involve an investment of around $250 million and employ from 300 to 500 people. At the time that BCWM was searching for a new site, it was under the one-man management of Dr. Christian Rosenthal, its major stockholder. He took direct authority over all siting decisions.

The 1972 Olympic Games in Munich caused a great deal of growth near BCWM's facility (today, over 40,000 people live within a one-mile radius of the existing plant). Expansion appeared to be out of the question. To complicate matters further for the firm, as the Olympic developments were proceeding, federal government officials discovered that Barlöcher did not have necessary licenses for its existing production facilities. They began putting pressure on the firm to adopt pollution-abatement measures.

This prompted Rosenthal to begin looking outside Munich for added capacity. BCWM was attracted to the Alsace region of France in 1973 by its cheap labor and direct access to foreign markets (through the planned Rhine-Rhone canal) and by significant incentives offered by the French government, which was attempting to lure foreign firms to economically depressed border regions.

The company created a French subsidiary and began discussions with officials in Saint Avold in the Moselle department, but environmental opposition there by local conservationists and farmers persuaded the company to focus on Marckolsheim, another Alsace city, where the local government was very receptive.

For construction of its lead stearate plant, Barlöcher needed two major approvals—a construction license and an authorization for "classified installations"—from the prefect (the local representative of France's central government). The prefect would be assisted in evaluating applications by central government agencies in Paris, notably the powerful Service des Mines (a technical agency connected with the Ministry of Industry) and the Ministry of the Environment. Local zoning was not an issue, given the fact that the local government was intent on finding firms to occupy its vacant industrial park.

Although the political responsibility for the licensing decision lay with the prefect, in fact the Service des Mines was the key actor. It controls the authorization of "classified installations." Before granting such an authorization, the Service prepares and negotiates with the firm involved "technical dossiers," which detail emission standards, monitoring requirements, and the like. Certain minimum standards, generally weaker than comparable U.S. standards, govern many production processes, but tighter controls can be imposed. Economic feasibility plays a key role in these negotiations. Other agencies and the public comment on these dossiers, but their role is advisory only.

The major environmental concerns for the Service des Mines in this case were air pollution, notably lead and cadmium emissions, water effluents and cooling, and noise. Local citizens and environmentalists, however, voiced greater concern about overall industrial expansion plans for the region, which included proposals for nuclear power plants, and about nature protection in an area that was still largely rural and agricultural.

Barlöcher conducted extensive preliminary negotiations with the Service des Mines, which was ultimately satisfied that the project would be safe and feasible. A formal application was then submitted in March 1974. Service officials say that the technical information compiled for the project made the dossier more exhaustive than any done previously. The basic principle guiding the negotiations was to use the most advanced technology available that would still be economically feasible.

The final approval, granted in June 1974, contained specific limits on lead emission that were tighter than existing French standards and ordered that water effluents be recycled rather than discharged. It also required automatic monitoring devices and an automatic system to shut the plant down if lead emissions exceeded allowable limits. Moreover, any changes in plans would require a new approval by the prefect. The Service des Mines and the company were pleased with the dossier and the controls imposed. Both felt that the dossier would convince local residents that the project posed no dangers.

But local citizens, irritated because they had been excluded by the company and central government from the secret negotiations, began to question the project. The Alsace region had a reputation for environmental awareness; ecological candidates had done quite well in local elections and better there than anywhere else in the 1974 presidential election.

The opposition came to a head at advisory public hearings on the project held in April and May 1974. Project opponents were well organized and presented a sound technical case that raised many questions about the project. The firm and the Service des Mines were ambiguous in responding to criticisms; for example, the potential for expanding the plant to produce dangerous cadmium stearates was never denied. Rosenthal refused to appear because, in his words, "balanced" public opinion could not be guaranteed.

Because of the hostile reception the project received, the local council and mayor voted against it. But the central government, through the prefect, ignored that vote and granted the necessary permits in late June. Construction was prevented, however, when environmentalists and other project opponents, including local farmers, took over the site. The occupation lasted several months, because public authorities decided not to intervene. The case had by then become a celebrated regional issue, and in the fall of 1974 it helped elect a majority of new municipal councillors sympathetic to environmental protection.

Barlöcher was still intent on pushing the project, but by early 1975 the central government was looking for a way out. The government wiggled free from the controversy by refusing to sell land in the industrial zone to Barlöcher, all the while maintaining that the environmental assessment and controls were perfectly adequate. Outraged, Dr. Rosenthal sued the French government for $2 million, but that suit was ultimately dismissed. Barlöcher then abandoned the Marckolsheim site and applied for permission to construct the facility in Kiel, West Germany. However, the company ran into trouble there, too, when the local council refused to sell needed land.

The firm next began negotiations with the city of Braunschweig, an industrial center of 270,000 in the southeast of Lower Saxony, West Germany. Braunschweig offered significant financial incentives for new industries as well as excellent transporation by air, water, and railroad. It also sported several excellent research institutions and a host of recreational opportunities, particularly the Harz and Elm mountains.

As in Marckolsheim and Kiel, Barlöcher needed local approval for the sale of land. The city council required BCWM to commission an "unbiased" report on the project, but at the same time it expedited necessary local building permits. The city council eventually gave its consent to sell land for the site contingent on a favorable environmental report and compliance with German federal laws. But the public, by now, was suspicious. When BCWM applied for necessary permits in February 1975, as the West German federal Clean Air Act required, over 10,000 objections were filed. Existing local industries also protested, fearing that the chemical plant would endanger their own expansion plans.

West German opponents organized into a local citizens' action group and recruited several respected scientists to evaluate the project. These experts compiled a decision guide that blasted the project for being too close to an already urbanized area. The guide claimed that the air in the Braunschweig area was already so polluted that additional lead emissions would cause irreparable harm to humans as well as to the soil and plants. Barlöcher attempted to refute these allegations—the firm pointed out, for example, that lead emissions would be only one-fifth as great as originally allowed by French authorities—but the tide of local opinion was clear.

In September 1975, Barlöcher surprised everyone by withdrawing its application in favor of a site in Lingen, a small town of 47,000 in the western part of Lower Saxony, after the city council there unanimously approved its building plan. Land-use problems in Lingen were considerably fewer than elsewhere, in large part because the city had located its industrial zone away from residential areas. It had also made significant infrastructure investments, notably in roads and waterways, and was a "preferred" industrial area that received federal government investment subsidies of up to 20 percent.

In Lingen, Barlöcher again ran into local environmental opposition. But this time the opposition was less well organized and did not present as solid a technical case as objectors in Marckolsheim and Braunschweig had. All necessary state permits were granted by June 1976. Opponents sued, but a lower court rejected their claims. Further appeals were impossible because of the threat that costs might be awarded against the opponents. (In most European countries, the losing party in a court case must not only pay the other side's court costs, but also the legal fees.) BCWM began construction on May 12, 1977, and production commenced in the fall of 1979.

Meanwhile, back in Munich, Barlöcher's original plant was under fire. Citizen's groups were attempting to shut that facility down on air-pollution grounds. Even though Barlöcher originally said it would transfer most of its lead production to Lingen, the company nevertheless applied for a production permit to continue operating in Munich. Despite opposition, that permit was granted by the city in 1978, with a proviso that lead production cease by 1980.

Still, citizens were not satisfied, and, based on monitoring data they produced at their own expense, they persuaded the Bavarian state government to monitor lead and cadmium emissions in the area. The group also succeeded in securing a court order overruling the municipal permit and directing the shutdown of several BCWM units. This action was reversed, however, when monitoring data from the state environmental protection agency showed that under normal conditions air-quality standards were not exceeded. More recently, cadmium effluents were found in a sewer, thus reagitating citizens' groups, who are demanding that the facility close.

At the beginning of BCWM's search for a new site, its management, particularly Rosenthal, adopted a hard-line negotiating attitude.

In its dispute with Munich authorities, the company threatened to sue if land-use plans were changed. The same approach was used in Marckolsheim, where Rosenthal attacked opponents as left-wing radicals. These statements backfired and created sympathy for those who opposed the plant. In Braunschweig, Rosenthal also personally intervened and again attacked opponents as radicals and communists. Early contact with the public and an open approach might have defused these controversies, but the company expressed little interest in dealing with anyone except government officials.

BCWM changed its approach in Lingen. The company's success there was at least partially attributable to a new corporate attitude. Until then, Rosenthal had generally left public relations to an outside advertising agency, presuming that a decent publicity leaflet would do the trick. However, the succession of defeats and potential problems in Lingen changed his mind.

To better its public relations, Barlöcher brought in a former manager of the Munich works, named Kraus, who had witnessed the earlier debacles. He immediately initiated a more open information policy and frequent meetings with citizens' groups. The company later breezed through a new licensing procedure for production of stearic acid with no formal objections at a public hearing.

This new openness met with some resistance within the firm. Moreover, Kraus felt he was called in only after all else failed. Thus, he resorted to maintaining his own informal "information" network to keep tabs on what other divisions in the company were doing. However, the company found that presenting sound technical information early in the planning and permitting process and responding to questions with solid data rather than diatribe could pay dividends.

The traditional role of governments in industrial siting in Europe has been similar to that in the United States. In the Barlöcher case, the governments' main goal was to attract industry. Only after that was accomplished were environmental considerations examined, usually under public pressure. Moreover, most governmental agencies, particularly at the local level, lacked technical expertise to evaluate adverse impacts adequately. Closed door negotiations exacerbated suspicions of citizens that their interests were not adequately represented.

The Barlöcher case is a good example of the siting process in transition. Just as the company was faced with new demands, so were the regulatory agencies involved. Perhaps more than anything, the agencies learned that a purely technical justification for a major project, however sound, does not necessarily ensure public acceptance.

Overall, the siting process in West Germany and France is marred by less overlap and fragmentation than in the United States. In France, the central government has final say on all large projects. In West Germany, all federal environmental standards, as well as land-use controls, are administered by state governments (locals do have a say, but not a veto). Moreover, the permitting process involves, by law, a greater degree of balancing among environmental and economic issues.

Although these attributes do have advantages for industry, they have also led to problems. For example, citizens generally have less say in the permitting process, and, because of limits on court actions, are more likely to rely on extra-legal means of opposition, such as disruptive demonstrations. Furthermore, while the regulatory process ostensibly involves a balancing process, neither country has developed tools to make that balancing anything but a political process.

The Barlöcher case is not the average European siting example. More frequently, the highly centralized French system, with its close collaboration between government agencies and industry, succeeds in getting facilities built. In two French case studies conducted for this project, plants were constructed expeditiously, although serious questions regarding environmental impacts remained unresolved.[38] All in all, however, environmental disputes are occurring in Europe with increasing frequency. Projects are delayed just as they sometimes are in the United States. Thus, even though Barlöcher did finally build its plant, the parallels with the Dow and SOHIO cases are striking.

STEAG[39]

Another German case study, this one focusing on the siting of a major coal-fired power plant, invites comparison with the difficulties experienced in siting big energy facilities in the United States.

Like the United States, West Germany experienced rapid economic growth during the 1960s, with a concomitant increase in the demand for electricity. Responding to projected needs, a government-controlled utility, called STEAG, based in the heavily industrialized Ruhr region, in the late 1960s began looking for a site for a major coal-burning power plant. Coal was the favored fuel under a national policy aimed at reducing reliance on foreign oil, and STEAG had direct interests in substantial coal reserves.

STEAG's original plan was to build two 350-megawatt units and one 600-megawatt unit to supply electricity to a regional distribution firm (which owned stock in STEAG). The second phase included a 700-megawatt plant worth almost $500 million, one of the largest projects of its kind at the time. Even in the mid-1960s, however, STEAG found that locating an acceptable site in the Ruhr area was no easy matter; many potential locations were ruled out because of preexisting high levels of pollution. As a result, STEAG concentrated its search on the fringes of the Ruhr area, focusing on established industrial zones. It finally settled on the municipality of Voerde in the heavily industrialized state of North Rhine-Westphalia. This site seemed particularly suitable because it was near coal pits, and the utility could obtain cooling water from the nearby Rhine River. The area had good transportation, was zoned for industrial use, and appeared to meet the existing federal guidelines for ambient-air quality.

The chosen site was in a rural area, home to many white-collar commuters who fled the crowding and pollution in nearby cities. But it was not these urban escapees who mounted opposition to STEAG's plan. Rather, a single person, a former engineer and businessman named Bassfeld, spent thousands of dollars of his own money fighting the project. Bassfeld's main complaint was that the ambient-air quality in the region was so bad that to permit further industrial development would endanger human health.

German environmental law is quite similar to U.S. law in ambient-air standards. Generally, new firms cannot locate in an area with dirty air, even though the proposed facilities would use the best available control technology, unless they agree to clean up existing plants. The plants already in these areas are protected from new regulations requiring better pollution control by a provision in federal law directing that in many instances companies must be paid compensation if their plants are forced to shut down or add controls. Just as in the United States, new plants bear the burden of improving air quality.

It is fair to say, however, that the majority of residents in the Voerde area did not support Bassfeld. In fact, labor unions in the region, where unemployment was high, came out strongly for the

THE PERMITTING-MAZE SYNDROME 97

plant and harassed Bassfeld by publishing his telephone number in a union publication. Several unions also threatened violence later in the dispute when Bassfeld said he was considering organizing a site occupation.

In the face of this resistance at the local level, spearheaded by the engineer Bassfeld, STEAG lobbied the local council and offered financial support for community needs. This tactic worked. The company secured local approval in 1967 for phase one of the project. In 1968, the state Industrial Supervisory Board issued initial approval for one 350-megawatt and one 600-megawatt unit. This initial phase came on-stream during 1970-71.

In 1973, STEAG applied to expand the facility by adding a 700-megawatt unit and increasing the capacity of one of the existing units. Considering the size of the project, approval was secured quickly. The project was announced in November 1973, hearings were held the following January, and interim permits were granted in September 1974.

But storm clouds were gathering. Over 2,200 objections to the project were filed during the administrative proceedings. People were becoming more environmentally aware, and, as in the United States, German environmental laws were being tightened. In 1974, the Bundestag (West Germany's parliament) enacted a federal pollution-control law. Soon thereafter, new air-pollution guidelines came into effect. These new ambient-air guidelines were the basis of a suit brought by two people objecting to the approval of STEAG's expansion.

The lower court sustained the objections and canceled the approval, ruling that construction was prohibited under federal law because proposed mitigation measures were inadequate for an area already violating federal air-quality guidelines. The national legislature reacted and pushed through a hotly contested change in the law that supposedly allowed the project to proceed.

But the appeals court confirmed the lower court result on different grounds that sidestepped the amended air law. The court said that meeting standards in the federal law did not mean that a permit had to be granted. Project proponents had to prove to the court's satisfaction that the public health and welfare would not be harmed by the project even if federal regulations were satisfied. The effect of that decision was to place the final destiny of every power plant in judicial hands.

This time the federal legislature was less enthusiastic about enacting special legislation to help STEAG, although there was a great deal of concern that the air laws, as interpreted by the courts, could thwart industrial development. Instead, the federal government created a $100-million fund to assist utilities in meeting pollution requirements.

STEAG's plans at Voerde were further affected when the company ran into opposition on another project—the construction of a power plant near the town of Bergkamen in the Ruhr area. The company agreed to pay several million dollars to Bergkamen itself and to an opposition citizens' action group from the town. The payments created a national outcry and charges of bribery. A lower court declared the payments illegal on grounds of public policy. Although the payments were eventually upheld on appeal, the incident tarnished STEAG's reputation across the country.

However, STEAG was bailed out for the Voerde project on February 17, 1978, when the highest federal administrative court reversed the lower court decisions interpreting the national air-pollution laws and dismissed the environmentalists' suit. It held that if any industry met the federal ambient-air standards, then that was probative evidence that a permit should be issued. The court also held that STEAG's plans to clean up emissions from a power plant several miles away could be counted as an "offset" against any new pollution.

Based on the high court decision, the project was quickly approved and operating permits were issued. But construction did not start immediately. STEAG's partner in the project, RWE (a large German utility), demanded a reapportioning of financial risks. It worried that a suit filed challenging the operating permits might lead to long delays or ultimate disapproval if the air-pollution laws changed; at that time, the federal parliament was considering several pieces of proposed legislation that would have tightened the existing air-pollution law. Some say RWE also delayed the project as a way of pointing out that it wanted clearer standards set out in federal environmental law so that, once they were met, a firm's project could not be challenged in court.

STEAG eventually agreed to RWE's financial demands, and construction began in 1981. However, a suit was filed challenging an operating permit that was required in addition to construction permits. This suit was dismissed by a lower court but is still being appealed. The Bundestag continues to debate changes in national air-pollution law, focusing on whether environmentally sensitive "non-degradation" areas should be identified and protected and on the details of a national offset policy designed to facilitate industrial growth in areas with dirty air.

STEAG's problems were caused in large part by uncertain and changing regulatory requirements. West Germany was at the time undergoing a major change in environmental protection, just as the United States had a few years earlier. In many areas, candidates from the Green or environmental parties were making significant electoral gains, particularly at the local level. To exacerbate matters, air-quality data were scarce or uncertain for the region around Voerde, so that companies like STEAG found it difficult to predict whether standards could be met and upheld in court.

From an organizational standpoint, STEAG was unprepared for controversy. The amount of environmental planning that actually went into its 1973 expansion was, by present-day standards, quite minimal. The head of the planning department, responsible for new construction work, planning, and environmental protection, had only one person to aid him in planning the expansion. A separate department has since been established for "environmental protection, industrial siting, and licensing." Since 1975, in compliance with federal law, STEAG has also employed an environmental safety officer, whose views must be heard whenever decisions are being made about new investments. In addition, the company has hired a landscape planner, whose job is to reduce negative visual impacts of STEAG's plants, an objective raised by some people early in the siting process.

The STEAG case did force the state and federal governments to focus on a range of important economic-environmental issues. How could industrial growth be permitted in areas with dirty air? Did current law discriminate in favor of existing, heavily polluting plants? These issues are still being debated in West Germany.

Unlike the Barlöcher case, in the STEAG dispute one man alone spearheaded the opposition. Bassfeld's persistence not only forced the company to adopt feasible environmental protection measures, it also helped focus national debate on the environmental impacts of industrial growth in areas that already had serious pollution problems.

On another level, however, Bassfeld represents the "extreme environmentalist" talked so much about these days. In truth, there are not many Bassfelds around, either in the United States or West Germany, but his actions illustrate the fact that a determined objector, acting alone, can use environmental laws to tie a huge project in knots. Bassfeld sought no compromise; thus no level of pollution control at the plant would satisfy him. He simply wanted the project killed. Dealing with people like Bassfeld is difficult, if not impossible, even for a project proponent willing to involve citizens early on in project planning.

European and U.S. Siting Parallels

The first thing that strikes the American reader about these foreign case studies is how much the disputes resemble those in the United States. Thus, despite West Germany's image of a well-oiled industrial machine, firms there must also grapple with an increasingly lengthy and strict permitting system full of uncertainties.

Extended judicial review can also be a problem. Moreover, although the United States currently appears to have a greater number of, and more complex, environmental protection laws, citizens in countries like West Germany are demanding that their governments catch up. The response is often to emulate U.S.-style controls. As one European governmental official observed: "America is the leading country in the world. If Americans like to sing and dance, everybody . . . is singing and dancing. The same is true with protest about nuclear power or pollution."[40] Environmental impact statements are becoming more widespread throughout Europe with the backing of the European Economic Community, and in West Germany an air-pollution "offset" policy is in the works.

Firms operating in Europe have to cope with a system of land-use controls far more pervasive than the system in the United States. Every square inch of West Germany, for example, is subject to strict land-use control pursuant to both state and local plans. In the United Kingdom, the government can deny any project on land-use grounds without compensation even if land-use plans specifically allow industrial use of a parcel.

As regulatory systems in Europe and elsewhere are tightened, environmental disputes, not surprisingly, are on the rise. While the two case studies in this chapter are not typical, there is little doubt that the United States does not have a corner on environmental conflict. A survey of hundreds of siting cases around the world concluded that:

> Environmental conflict has been spreading internationally as well as inter-

regionally, with most battles in the early part of the decade concentrated in heavily industrialized parts of the U.S., U.K., Germany, the Netherlands and Japan, but later emerging in virtually every other developed nation—and more recently existing, albeit sporadically, in rapidly industrializing nations such as Brazil.[41]

When they do erupt, environmental disputes in Europe are often more violent than those in the United States. A number of knowledgeable observers attribute this to the way European citizens are shut out of the decision-making process, with negotiations between government and industry carried out in private.[42] Citizens may have a chance to comment, but the feeling is widespread that the final result is predetermined. The feeling of helplessness is made worse because Europeans generally have limited access to courts to settle siting disputes, either because of the ways laws are written or because of the specter of having to pay court costs and attorney's fees for the opposing party.

The European cases also invite comparisons between the way domestic and foreign firms plan big projects and evaluate potential environmental impacts. Perhaps because U.S. firms had to face a panoply of new environmental laws and demands earlier than did companies abroad, many now are more sophisticated than foreign firms in dealing with the public and conducting project planning in the open. As one Dow official with wide experience in siting plants around the world remarked, "We work with the government of whatever country we are in and do what is required of us."[43]

There is evidence that corporate project planning in Europe, particularly, is improving as it has in the United States. As one German commentator observed, "This is a period of learning in Germany as well."[44] Barlöcher certainly learned that a more open planning process is essential and that secondary growth impacts are increasingly important in siting decisions. The same can be said of government regulators. They are realizing that a "we know best" attitude is increasingly unacceptable.

One attractive attribute of the European approach, at least to industry, is that government and business work cooperatively. Although this attitude of close cooperation can backfire, as it did for Barlöcher and STEAG, the general absence of an adversarial atmosphere sometimes helps in working out solutions to difficult pollution-control or land-use problems that might otherwise lead to delay or litigation.

In fact, European firms seem more willing than U.S. firms to work with government early in the planning process in some cases, perhaps because negotiations take place in secret. That is all very well for reaching quick agreement, but it is the down side of the "partnership" between business and government from the perspective of excluded citizens' groups. Striking a balance between fruitful, but armslength, cooperation and unseemly collusion that excludes the public poses a great challenge.

There also appears to be less interagency overlap and conflict in most European regulatory systems than in the United States, mainly because of the way governments are organized. In West Germany, a single state board is responsible for virtually all environmental permits (except those relating to water), and there are time limits on decision-making. In France, the central government, through its prefects, has absolute authority over most projects, thus "eliminating" conflicts. Of course, localities there have little real say over big projects. The United Kingdom has a system that preserves local control, particularly over land use, but allows the central government to "call in" controversial applications for decision. However, despite this freedom from a maze of regulatory agencies, the environmental and land-use control systems in Europe remain complex and time-consuming.

Summary

The foreign case studies reveal several important points about the foreign experience:
- The United States does not have a corner on environmental-industrial disputes.
- European nations are emulating the U.S. approach to environmental regulation in some important respects and already have stronger land-use laws.
- European firms and governments are experiencing the same "teething pains" in dealing with new environmental and land-use laws.
- European corporations and governments are not as advanced as U.S. firms in assessing the environmental impacts of a project early on nor as adept at involving citizens in that process.

A good deal can be learned from the foreign experience, but much of it is not how they do it better. Contrary to popular wisdom, European countries such as West Germany and the United Kingdom have land-use controls far stronger than those in the United States, and it often takes a comparable amount of time to site a plant in Europe,

Japan, or elsewhere as here. Furthermore, the lack of direct citizen participation in many siting decisions has helped fan bitter opposition to some projects. Siting disputes in Europe and Japan appear to be increasing and are no less intense than in the United States.

The European experience shows that regulatory problems are unlikely to be solved by retreating to the "old way" of doing things— through less citizen involvement and more closed-door negotiating between government and industry, through a weakening of environmental laws or ignoring of environmental concerns. Those methods do not even work well in Europe any longer. In short, regulators and industry can't "go home again" to the seemingly simpler, easier way of siting new facilities that prevailed well into the 1960s.

NO NEED FOR REFORM?

The evidence that environmental and land-use regulations are not the impediments to industry that they are often portrayed to be is substantial. Yet there are compelling reasons to avoid complacency. The United States cannot ignore the significant costs that environmental protection entails, just as it cannot ignore the important benefits. Society needs to be concerned about the handful of industries that do appear to have suffered because of compliance costs. Furthermore, although economists generally agree that environmental restrictions do not have a large direct impact on the economy, they are still uncertain about indirect impacts on productivity, innovation, and inflation.

There is no real way to measure the extent of the "stillborn" projects phenomenon noted earlier, although the theme recurs frequently in discussions with industry representatives.[45] Finally, both environmentalists and businesspeople ask whether the complex regulatory system actually provides the protections offered in theory by lengthy reviews.

REFERENCES

1. Robert Healy and John Rosenberg discuss the growth in state land-use controls in *Land Use and the States* (Baltimore, Md.: Johns Hopkins University Press, 1980); see also Fred Bosselman and David Callies, *The Quiet Revolution in Land Use Controls* (Washington, D.C.: U.S. Government Printing Office, 1972); Nelson Rosenbaum, *Land Use and the Legislatures* (Washington, D.C.: Urban Institute, 1976).

2. As of 1979, 28 states had adopted "little NEPAs," according to Council on Environmental Quality, *Environmental Quality—1979* (Washington, D.C.: U.S.

104 PART II: PROBING THE HEADLINE CONCERNS

Government Printing Office, 1979).

3. Three of the more well known laws are from Florida, Wyoming, and California. The Florida Electrical Power Plant Siting Act, F.S.A. sec. 403.501 *et seq.* (1979), is described in Wade Hopping, "An Industrial Siting Act That Works: The Florida Electrical Power Plant Siting Act," a paper presented at the Eighth Annual American Bar Association Conference on the Environment, Airlie House, Warrenton, Virginia (May 4-5, 1979). Jack Van Baalen analyzes Wyoming's siting law, the Industrial Development Information and Siting Act, W.S.A. sec. 35-202.75 *et* seq. (1975), in "Industrial Siting Legislation: The Wyoming Industrial Development Information and Siting Act—Advance or Retreat?," *Land and Water Law Review*, vol. 11 (1976), p. 27. California's Warren-Almquist Resources Conservation and Development Act, Cal. Pub. Res. Code, sec. 25000 *et seq.* (1977), is described in J. Asperger, "California's Energy Commission: Illusions of a One Stop Power Plant Siting Agency," *UCLA Law Review*, vol. 24 (1977), p. 1313.

4. The boom in local land-use laws and attendant permitting issues is discussed in Fred Bosselman, Duane Feurer, and Charles Siemon, *The Permit Explosion: Coordination of the Proliferation* (Washington, D.C.: Urban Land Institute, 1976); and in John H. Noble, John S. Banta, and John S. Rosenberg, eds., *Groping through the Maze: Foreign Experience Applied to the U.S. Problem of Coordinating Development Controls* (Washington, D.C.: The Conservation Foundation, 1977).

5. Gary Knight, Speech to a National Air Quality Conference, San Francisco (January 15, 1979).

6. Michael Heseltine, quoted in John Blake, "Instant Justice," *Town and Country Planning* (June 1980), p. 180.

7. John Connell, "Perspectives on the Regulatory Maze," presented at a Workshop on Governmental Permit Simplification, Coordination, and Streamlining, sponsored by the Hawaii Department of Planning and Economic Development, Honolulu, Hawaii (July 7, 1978).

8. Pat Choate, *As Time Goes By: The Costs and Consequences of Delay* (Washington, D.C.: Academy for Contemporary Problems, 1980). Without defining the term *delay,* Choate discusses a variety of possible ill effects of increased time to complete projects. See also Gunther S. Schramm, "Anatomy of the Cost of Time," in Gunther S. Schramm, ed., *The Value of Time in Environmental Decisions and Processes* (Ann Arbor, Mich.: University of Michigan Press, 1979), p. 11.

9. Choate, *As Time Goes By*; Schramm, "Anatomy of the Cost of Time."

10. John Pinkerton, "Experience of the Forest Products Industry in Obtaining Air Quality Permits for New and Modified Sources," *Journal of the Air Pollution Control Association*, vol. 31, no. 11 (November 1981), p. 1163.

11. John Schark quoted in Lawrence Mosher, "The Clean Air You're Breathing May Cost Hundreds of Billions of Dollars," *National Journal* (October 10, 1981), p. 1816. For a further discussion of this point, see Paul Portney, "The Macroeconomic Impacts of Federal Environmental Legislation," in Henry Peskin, Paul Portney, and Allen V. Kneese, *Environmental Regulation and the U.S. Economy* (Baltimore, Md.: Johns Hopkins University Press, 1981).

12. Thomas Gladwin, "Patterns of Environmental Conflict over Industrial Facilities in the United States, 1970-78," *Natural Resources Journal*, vol. 20 (April

THE PERMITTING-MAZE SYNDROME 105

1980), p. 244.

13. The results of this Mid-Continent Oil and Gas Association survey are reproduced in American Petroleum Institute, *Major Legislative and Regulatory Impediments to Conventional and Synthetic Fuel Energy Development* (Washington, D.C.: American Petroleum Institute, 1980), p. 54.

14. See James Mahoney and Barbara Goldsmith, *Effects of the Clean Air Act on Industrial Planning and Development*, vol. 1 (Washington, D.C.: The Business Roundtable and Environmental Research and Technology, Inc., 1981). According to Goldsmith, this study initially reviewed approximately 300 cases cited by industry and subsequently focused in detail on 92. (Telephone conversation with Barbara Goldsmith, various dates in 1983.)

15. U.S. Department of Commerce, Bureau of the Census, *Statistical Abstract of the United States: 1981* (Washington, D.C.: U.S. Government Printing Office, 1981), p. 777. These figures are from the census of manufacturers, which was last completed for the year 1977. Figures for 1982 are not yet available. A company operating at more than one location is required to file a separate report for each site.

16. See Patrick F. Mason and Donald J. Skinner, "Raking in the Chips," *Planning* (November 1981), p 10.

17. U.S. Department of Commerce, *Statistical Abstract: 1981*, pp. 788-89.

18. "Trimming Down to Revive the Growth Pace," *Business Week* (June 1, 1981), p. 87.

19. See Nicholas Yost, "Environmental Regulation—Myths, Realities and Reform," *The Environmental Professional*, vol. 1 (1979), p. 257.

20. American Petroleum Institute, *Major Legislative and Regulatory Impediments*, p. 54.

21. Office of Planning and Resource Management, *Analysis of New Source Review (NSR) Permitting Experience* (Washington, D.C.: U.S. Environmental Protection Agency, 1982).

22. Ibid., p. 11.

23. Ibid., pp. 5, 8.

24. John Pinkerton, *Experience of the Forest Products Industry in Obtaining Air Quality Permits for New and Modified Sources* (New York: National Council of the Paper Industry for Air and Stream Improvement, Inc., 1981). Pinkerton's study, perhaps the most thoughtful and insightful evaluation of permitting problems prepared by industry in recent years, is summarized in the *Journal of the Air Pollution Control Association*, vol. 31, no. 11 (November 1981), p. 1163. Several other studies generally concur with the results of Pinkerton's survey. See Arthur D. Little, Inc., *The Effects of PSD on Industrial Development* (New York: The Business Roundtable, 1980); Environmental Research and Technology, *The Impact of Air Quality Permits Procedures on Industrial Planning and Development* (New York: Business Roundtable, 1980); and Dames & Moore, *An Investigation of the PSD and Emission Offset Permitting Processes* (Washington, D.C.: National Commission on Air Quality, 1980). For a contrast with the Pinkerton study, see the American Petroleum Institute, *Major Legislative and Regulatory Impediments*, which seems to advance the notion that anything that slows immediate development of energy resources constitutes an impediment and unwarranted delay.

25. If a large project is being built in an area that has air cleaner than national standards, the Clean Air Act requires the relevant air-quality agency to undertake a review and to impose measures to prevent significant deterioration.

26. This example is a summary of the author's actual experience in representing a utility in private law practice.

27. Urban Systems Research and Engineering, "Causes of Delay in Development of Major Industrial Projects," unpublished report for the Council on Environmental Quality (1982).

28. Marc Messing, "Reasons for Delay in Powerplant Licensing and Construction" (Washington, D.C.: Environmental Policy Institute, 1978). Messing recognized that the Federal Power Commission (FPC) data have two serious limitations: first, they represent information compiled by the FPC based on surveys of the electric utility industry, and certain aspects of the data suggest an accompanying bias. Second, the data are based on contributing causes, rather than on single or most important causes, and therefore any number of "causes" may be cited for the delay of a single power plant. Messing's study has been criticized by industry. George Gleason of the American Nuclear Energy Council, in testifying before the Subcommittee on Energy and Power of the House Committee on Interstate and Foreign Commerce on July 18, 1978, noted that the FPC data were internally contradictory and that the most frequently cited cause-of-delay category in the fourth quarter of 1976 was "not reported." However, the weight of Gleason's criticisms was directed toward conclusions regarding *nuclear* power plants, not fossil-fueled ones. Even so, Gleason conceded that "the nuclear industry recognizes licensing is not the sole cause of delay."

29. National Electric Reliability Council, *9th Annual Review of Overall Reliability and Adequacy of the North American Bulk Power Systems* (Princeton, N.J.: National Electric Reliability Council, 1979).

30. National Coal Association, *1979 Survey of Electric Utility Capacity Additions 1979-1988* (Washington, D.C.: National Coal Association, 1979), p. 17.

31. Mahoney and Goldsmith, in *Effects of the Clean Air Act*, pp. 5-15, concede that their sample is biased.

32. Contrast Mahoney and Goldsmith, *Effects of the Clean Air Act*, with American Petroleum Institute, *Major Legislative and Regulatory Impediments*. The U.S. EPA later reviewed the Mahoney and Goldsmith document and issued a scathing rebuttal. Mahoney and Goldsmith responded in a December 1, 1981, letter to Representative Henry Waxman (D.-Calif.), chairman of the U.S. Congress, House Energy and Commerce Committee, Subcommittee on Health and the Environment.

33. Office of Planning and Resource Management, *Analysis of New Source Review Permitting Experience*, p. 1.

34. See Gladwin, "Patterns of Environmental Conflict."

35. Interview with Jack Jones, Dow Chemical Company, in Walnut Creek, California, August 1979.

36. As a complement to The Conservation Foundation's U.S. research, the International Institute for Environment and Society (IIES), with the assistance of a French organization, Fondation pour le Cadre de Vie, conducted the major siting case studies as part of an effort to see what could be learned from the European

THE PERMITTING-MAZE SYNDROME 107

experience. Two of those case studies are presented in summarized fashion in this section. Longer versions are available from The Conservation Foundation.The Conservation Foundation augmented this work by drawing from the extensive project files, articles, and books it had produced as part of its International Comparative Land-Use Project conducted from 1974 to 1980 in seven nations (the United Kingdom, France, West Germany, Austraiia, Japan, Israel, and Mexico). Conservation Foundation staff studied several industrial siting disputes in those countries as part of that effort. That project was funded by the German Marshall Fund of the United States. Major publications and articles included:
- Fred P. Bosselman, *In the Wake of the Tourist: Managing Special Places in Eight Countries* (Washington, D.C.: The Conservation Foundation, 1978).
- Christopher J. Duerksen, "Dobry on Development Control in England," *Real Property, Probate and Trust Journal*, vol. 2, no. 2 (Summer 1976).
- George Lefcoe, *Land Development in Crowded Places: Lessons from Abroad* (Washington, D.C.: The Conservation Foundation, 1979).
- Noble, Banta, and Rosenberg, *Groping through the Maze*.
- Richard J. Roddewig, *Green Bans: The Birth of Australian Environmental Politics* (Montclair, N.J.: Allanheld, Osmun & Company, 1978).

37. This Barlöcher Chemical case study is a summarized version of one researched and written by Gabriele Knödgen of IIES and Michael Pollack, with the assistance of Ernst Brey. The longer version is available from The Conservation Foundation along with a detailed description of the French and German environmental regulatory systems.The other German case study, also prepared by Knödgen, examined the development of a large chemical complex by the giant Bayer firm in an agricultural area on the lower reaches of the Elbe River, near Hamburg. Bayer began searching for a site for this facility in 1969 and eventually settled on the Elbe River location. Site preparation began in 1972, but opposition by local citizens and environmentalists, who filed lawsuits, held up the project for over five years and resulted in plans for portions of the complex being abandoned. Key environmental issues included water quality and the cumulative impact of large industrial developments in an agricultural area. Bayer finally obtained all necessary permits and began construction in 1978. However, some local citizens, headed by fishermen who claim the plant is fouling the river with heavy metals, continue to oppose its operation.

38. The other French cases focused on a Shell polyvinyl chloride (PVC) plant and another company's fiberglass-production facility. The Shell affair provides a fascinating illustration of the centralized French siting system at work. There, the local government vigorously opposed the site chosen for the PVC plant in an industrialized area that was already beset by serious pollution problems. Despite this opposition and a delay of over two years, Shell succeeded in 1980 in getting needed approvals from the central government, which was firmly committed to the project because of the large investment involved.The fiberglass plant case was quite different. There, the main issue was retroactive application of pollution-control standards to a facility that began operation in 1972 without formal government approval. Negotiations between the firm—an old, respected company—and the central government over pollution controls took place from 1975 to 1981 while emissions from the plant continued. Little citizen outcry was heard during that

period. Finally, in 1981 a permit was granted defining acceptable emission levels from various sources within the plant. The case illustrates the great value that the French system places on having quiet negotiations rather than confrontations whenever possible.

39. This is a summarized version of a longer case study researched and written by Gabriele Knödgen. The longer version is available from The Conservation Foundation.

40. Ramon Leanato Marsal, Ministry of Industry and Energy, Spain, quoted in H. Jeffrey Leonard, "Environmental Regulations, Multinational Corporations and Industrial Development in the 1980s," *Habitat International*, vol. 6, no. 3 (1982), p. 9.

41. Thomas Gladwin and Ingo Walter, *Environmental Conflict and Multinational Enterprise* (New York: New York University School of Business Administration, 1979), p. 912.

42. Discussions with Thomas Gladwin, various dates in 1981-82.

43. Telephone interview with Ray Brubaker, Dow Chemical Company, Sao Paulo, Brazil, April 23, 1980.

44. Discussion with Gabriele Knödgen with Conservation Foundation staff in Berlin, October 1981.

45. Several industry officials made this point at conferences on the role of environmental and land-use regulations in industrial siting held by The Conservation Foundation in Washington, D.C., in June 1979 and San Francisco in May 1983.

Chapter 5

Attempts To Improve the Rules: Reforms of the 1970s

Before the ink was dry on the panoply of new environmental laws and regulations enacted in the early 1970s, people began to see shortcomings. The multiple-permit problem discussed in the last chapter was the first to receive attention. Books were written, conferences held, and new laws passed to deal with overlapping agency jurisdictions, contradictory permit requirements, and the like.[1] Streamlining and coordination were watchwords.

Later, frustration with the regulatory system led to calls for more drastic changes. Most of these proposals were marked by the almost timeless tendency to seek quick and simple solutions, to cut corners in the existing process in an effort to speed things up. But cutting corners or short-circuiting the existing process can cause serious problems. Ignoring environmental impacts or citizens' concerns might help speed things, but other values might be jettisoned and different problems created. The existing process is complex because environmental problems are difficult, and values like public health and quality of life compete with the desire for jobs and industrial growth.

The early attempts at reform teach some valuable lessons about the pitfalls in revamping the environmental regulatory system.

STREAMLINING THE PROCESS: ONE-STOP PERMITTING AND CONSOLIDATION

One highly touted reform idea of the last decade, something akin to a Seven-Eleven convenience store for environmental and land-use permits, would have a project proponent go to a single office for all necessary federal, state, and local approvals. A study of energy-facility siting summed up the arguments in favor of this type of reform as follows:

> According to proponents, the virtues of the one-stop process are that it is more efficient and less time consuming. It forces all the regulatory agencies to discuss and resolve issues between themselves and other parties to the

process. Unlike other review processes, where state agencies duplicate one another and already debated issues may be reopened, one-stop resolves issues in a complete and final fashion. Presumably, it increases the scope of considerations by forcing the interaction of all the parties to a decision.[2]

How did attempts at one-stop permitting work in practice? The definitive study on permit coordination concluded: "One-stop shopping—the delegation of all authority over land-use and environmental issues to a single 'czar'—cannot, realistically speaking, be accomplished. The issues are too complex; our political institutions too varied."[3]

There were also less ambitious attempts to solve the multiple-permit problem that varied from state to state. Most, however, relied on such improvements as master application forms for a project and consolidated state agency hearings to speed the process. The results have been disappointing, but the experience has been instructive.

In 1973, the Washington state legislature enacted an Environmental Coordination Procedure Act designed to untangle the permit web.[4] In this system, a master application is filed with the state Department of Ecology. The department circulates the application to relevant state agencies, which identify any permits needed and notify the applicant. At that point, the applicant can either obtain the requisite permits individually or can return all applications to the department, which will hold a consolidated hearing on behalf of all state agencies involved. Affected local governments have the option of joining in the consolidated hearings. After this hearing, permits are issued or denied by the individual state agencies and local governments.

Most project developers seek permits individually rather than through the system's "streamlined" proceedings. They say that the system has simply added another layer of bureaucracy and that the coordinated procedure takes longer than applying discretely to each agency. They also point out that local governments are not required to participate in the process.

The federal EPA also undertook a program to streamline its permitting system for five major programs.[5] The howls from several affected industries were hardly ones of delight. Although the new provisions called for consolidated hearings, they did not require that permits be issued at the same time, nor did they consolidate judicial review of the various permits once issued. Moreover, states could not be forced to adopt the federal EPA procedures. The coordinated system was virtually unused and was recently "decoordinated."

Beyond one-stop permitting to solve regulatory problems, other, more ambitious efforts were devised by some states to consolidate permitting agencies into a single body with responsibilities for all air, water, land-use, and other environmentally related programs and laws.

This approach was adopted by the central government in the United Kingdom when it created the massive Department of the Environment to oversee national land-use planning regulation, pollution controls, countryside and historic building conservation, transportation, and housing. A White Paper explained the underlying rationale: "Because these functions interact, and because they give rise to acute and conflicting requirements, a new form of organization is needed at the center of the administrative system."[6]

According to a 1978 report for the state of Maine, 15 states have adopted a variation of the British approach, consolidating pollution-control activities and natural-resource management into a single entity.[7] An equal number have lodged all pollution-control functions in one agency or department, and many have consolidated permit systems for major facilities, particularly power plants.

Consolidation raises many difficult issues. Which agencies should be consolidated and which left alone? What techniques should be used to fuse discrete agencies into a new working body? What sort of internal organization should be created? The experience with consolidation is mixed, indicating that these kinds of issues were not well addressed in the first wave of reform, and that problems formerly experienced *among* agencies are sometimes still felt *within* one new agency.

The Department of the Environment in the United Kingdom exemplifies coordination pains within a new consolidated agency. Originally, the government's Department of Transportation was brought under the environment umbrella in hopes of getting highway planners to work more closely with other planning agencies and local governments. However, transportation officials continued to operate autonomously, and eventually political pressures led the government to reestablish a separate transportation department.

In some U.S. cases, key agencies or permits have been left out of a reorganization. At the federal level, the EPA ostensibly has complete authority over water quality and pollution control. However, industries building in coastal areas know that if they intend to dredge and fill in a wetland, a permit is required from the U.S. Army Corps of Engineers.

Similarly, state consolidation efforts typically split environmental and land-use duties. Colorado's Department of Natural Resources oversees activities such as mining and is the state's development coordination agency for major energy and industrial projects. But the Department of Public Health issues air- and water-pollution permits.

And, of course, these consolidation efforts do little to improve coordination among various levels of government. In Colorado and many other states, air-emission permits must be secured from *both* the federal and state governments. Even in states that have been delegated federal authority, the permitting agency cannot ignore federal standards. Moreover, few consolidated resource agencies in states have any authority to issue local land-use permits.

Georgia's Consolidation Attempt

Another example of agency consolidation, and perhaps the most successful experiment to date, is the Environmental Protection Division (EPD) of the Georgia Department of Natural Resources (DNR). Because it appears to be working better than most other consolidation schemes and has the support of both industry and the conservation community, the experience is worth looking at in some detail.[8]

While he was governor of Georgia, Jimmy Carter pushed through a comprehensive reorganization of state government. A key element consolidated all state environmental-permit programs under the director of a newly created division, the EPD of the DNR. Unlike reorganization efforts in other states, this initiative placed virtually all important state environmental permits under EPD's wing, from air quality to groundwater allocation and from erosion to sedimentation control.[9]

Although the DNR is a politically appointed body, the EPD and its director (appointed by DNR) have independent status. This vesting of such great power in an independent administrator, which acts to shield the division from political influence, caused worry, particularly in industry. To allay concern, a compromise was reached: the director's decisions can be appealed to an independent hearing officer whose ruling is then reviewed by a five-member panel of the DNR. This procedure helps insulate the director from political meddling and, at the same time, avoids attempts to have the governor intervene directly in the regulatory process.

Industry supported the legislation because it offered few opportunities to appeal or challenge EPD decisions, thus reducing uncertainty due to litigation. Challenges must be filed within 30 days of a final ruling.

Conservationists, too, found the reorganization to their liking. It promised to provide a comprehensive review of projects (Georgia has no environmental impact statement requirement) and created a highly professional organization.

The Georgia approach does not seem particularly revolutionary, aside from the fact that EPD does have broad permitting authority. But the results are impressive. In announcing a $300 million pulp-and-paper mill near Augusta, Georgia, Kimberly-Clark said one of the reasons it chose the state was its one-stop permitting. State officials say that Miller Brewing chose a Georgia site for a plant to serve the Southeast over sites in Tennessee, South Carolina, and Florida because only Georgia could guarantee that permits would be reviewed to meet the company's planning deadlines. One industry official summed up Georgia's attraction: "What's unacceptable is different answers from different agencies or levels. Georgia doesn't have that problem; it can give just one answer."

The Georgia system works for several reasons:

- *Strong administrative support.* EPD has received strong support in terms of money and personnel from the governor's office and the state legislature. One observer said, "Governor Busbee's implementation and support made it fly." (Busbee was Carter's successor.) The state attorney general has also contributed by assigning a lead and backup attorney for each law and program. This not only gives EPD solid legal counsel but has been helpful to industry in explaining and clarifying federal and state regulations.
- *Federal delegation of environmental reviews.* The EPD has aggressively pursued and obtained delegation of major federal environmental permitting programs under the Clean Air and Clean Water acts. It also has signed an agreement with the regional EPA office whereby the EPA agreed to review permits expeditiously—within ten days—when its review is required by law. In addition, EPD has been particularly careful to ensure that all Georgia laws are consistent with federal legislation. These efforts have cut duplicative federal-state permit requirements greatly.
- *Good communication between EPD and applicants.* Most major

industries locating in Georgia first contact the Department of Industry and Trade (DIT) for advice on sites, permits, and the like. DIT and EPD maintain a good, informal working relationship, and industries are usually steered immediately to EPD if there appear to be potential environmental problems. Industry and state officials say this informal referral has helped avoid the selection of sites that lack a fighting chance of success. EPD has also organized a special in-house task force for dealing with major industries. Headed by the EPD director and four hand-picked technical experts from the division who have intimate knowledge of state regulations and the applicant's industry, the task force meets early with industry representatives to explain laws, answer questions regarding specific industrial processes, and so forth. The task force later reviews permit applications submitted by the project proponent. The EPD director spends 15 to 20 percent of his time working with the task force and believes it is well spent. "It prevents most problems from occurring—we've coaxed industry away from inappropriate sites—and it [my participation] impresses on my staff that the project should be processed expeditiously."

- *High degree of consensus on environmental issues.* Georgia is seemingly blessed with an amazing degree of consensus about industrial growth and environmental protection, one not found in many other states. Although groups like the Sierra Club and other national organizations are active in the state, the Georgia Conservancy is the dominant environmental organization, and it sets the tone on environmental issues in the state. The present governor is a member, and EPD has an excellent working relationship with the Conservancy.
- *A respected, independent director.* The director of EPD receives high marks from all sides for independence and integrity. Some feel the system will not work when he leaves. Although EPD frequently meets and works with industry, no one suggests that the process is "wired" or closed to the public. Some environmentalists view the director as proindustry. Nevertheless, they say he is fair and follows the law.

Thus, the interesting combination of a statutory reorganization, informal procedures, and a highly respected director have made the Georgia one-stop permitting system work.

Are there no shortcomings? Even EPD's director concedes that the process is not geared to evaluate local secondary-growth impacts of major projects; thus, local governments often are left to their own devices to analyze and cope with the indirect impacts of major projects. The problem is exacerbated by Georgia's almost complete lack of zoning. Unzoned areas therefore lack leverage to bargain with new industry over mitigation measures. The director says the state needs to do a better job of providing advice and technical assistance.

Some critics say that the permitting process works in part because people are just not that interested in environmental issues and the state agencies are not about to rock the boat. One state official commented, "Here most people join the Sierra Club to go on hikes or have fun, not to be activists."

It is not likely that many states can enact a comprehensive reorganization that places tremendous power in a nonelected, independent official as Georgia has done. On the other hand, the Georgia system has several features that can readily be adopted elsewhere. The informal liaison between EPD and DIT and the industrial-technical task force are two excellent examples.

Use of comprehensive consolidation to overcome the multiple-permit problem may work in some places, but in most jurisdictions comprehensive reform such as Georgia accomplished will be difficult. Entrenched bureaucracies, industry and public skepticism, and local opposition will all work against reform on this scale. Moreover, even if limited consolidation can be accomplished, it will not solve all the problems of the siting process.

TRUNCATING THE PROCESS: IGNORING CERTAIN LAWS, GOVERNMENTS, AND CITIZENS

As reformers began to see the limitations of streamlining, some decided the answer to the proliferation of permits, hearings, and agencies required a harder line: get bothersome citizens, public-interest organizations, laws, and governments *out* of the process altogether, particularly when a major, highly visible project is at stake.

The theory behind removing citizens and public-interest organizations from the process, for example, is that things would go more smoothly if outside people and groups were not allowed to tie things up in hearings required by environmental laws and if opportunities to appeal decisions were reduced.

Under President Reagan, the federal government has attempted to reduce citizen involvement in the regulatory process. The Office of Management and Budget (OMB) made a concerted effort to cut back on opportunities for citizen participation in the so-called Section 106 process, established pursuant to the National Historic Preservation Act of 1966 to evaluate the impacts of federally funded or approved projects on historic areas and landmarks.[10] In addition, the Department of the Interior (DOI) has been trying to reduce "burdensome and unnecessary" public participation rules that govern management of federal public-domain lands.[11] Moreover, DOI has been seriously considering restricting the rights of private citizens to challenge departmental decisions; challenges would be allowed only by parties with "property" interests adversely affected by a decision. Thus, an oil or timber company doing business on public lands would have standing, but a hunter or fisherman who used the same land would not.

Other attempts at truncation have relied on establishing one agency or board to oversee the siting process. Many of the earliest of these reforms proposed giving a "super-siting" agency within a state government all authority over major industrial and energy-related developments to counter both the state regulatory bodies that were rapidly proliferating and local governments that were increasingly asserting power over such projects (particularly with land-use controls).

One of the first and most publicized initiatives was promoted by an American Bar Association special committee on environmental law in the early 1970s. That committee undertook a comprehensive three-year study of legal reforms to improve the decision-making process for industrial-site selection. In its final report, the committee, citing regulatory overlap and confusion at the local level and among state agencies, recommended that all authority over major industrial projects be moved from local governments to a state industrial-siting council.[12]

The committee's proposals generated much debate but were never formally approved by the American Bar Association. However, they did serve as models for several state siting laws, and the seeds planted by the report continue to sprout. The centerpiece of a recent study, prepared for the American Petroleum Institute on coastal energy-facility siting, was a state energy review board with power to "overrule a state or local agency's final decision" to stop a major energy project.[13] State siting councils have also been promoted as the answer to local opposition to siting hazardous-waste disposal facilities.[14]

President Carter proposed adapting this reform at the federal level. He advanced legislation that would have authorized a new federal energy mobilization board (EMB) to override state and local objections to priority energy projects.[15] The premise was that state and local governments were needlessly delaying projects. As one EPA oficial reportedly said, "Most permitting is state and local permitting, and so is most permitting delay."[16] Later versions of the EMB legislation would have allowed federal intervention only after state and local governments failed to act in a specified period.[17]

The truncation approach also seems alive and well in the Reagan Administration.[18] The National Oceanic and Atmospheric Administration in May 1981 proposed a rule change under the federal Coastal Zone Management Act designed to strip states of the authority to analyze offshore oil and gas pre-lease programs for their consistency with the states' coastal management plans.[19] States vehemently objected to that move, arguing that they have a right to analyze onshore economic and environmental effects of offshore drilling since the states would shoulder such impacts. Under fire from Congress, the Administration later withdrew the proposed regulations.

Similarly, DOI attempted to reduce the say states have in the leasing of coal deposits on federal land within their borders. During the 1970s, the department established consultative regional councils, composed of federal, state, and local officials, to set levels of leasing for future coal development consistent with state and local governments' abilities to withstand the resulting social and economic impact. Although that process appeared to be working well, Interior Secretary James Watt changed it so that DOI is no longer bound by the recommendations of the regional panels.

Governors of 13 western states protested vigorously and have threatened to seek changes in federal mining law to require greater consultation. An aide to Governor Richard Lamm of Colorado observed, "Now we realize that while he's [Secretary Watt's] been dancing us around the ballroom, the Interior Department has been working all the time to dismantle previous areas where the states had real input."

Interior has taken other steps to reduce state and local input and ignore the impacts of major projects. DOI abolished the Office of Special Projects within the Bureau of Land Management, which had shown great promise in making the review process for major developments (such as pipelines) on public lands both more efficient and effective.

Congress, too, has shown no shyness about enacting legislation to bypass state and local permitting authorities to force construction of projects it really wants. Two prominent examples are the trans-Alaskan oil pipeline and the Tellico Dam. But the tarnished records of these two cases provide evidence that, if similar exemptions were to become the norm, the system would bog down in legislative, rather than administrative, red tape.

Occasionally, however, there may be a project of such critical national importance that a truncated permitting system is merited. In that case, a few state siting acts might offer useful models for the federal government. Florida and Washington State, in particular, have adopted laws that attempt to balance speeding up the permitting process with recognizing the legitimate regulatory interests of various government agencies.

Florida

Florida has an Electrical Power Plant Siting Act (PPSA) that has, over time, won support from many affected parties.[20] The PPSA requires that any utility company planning to construct a power plant file a comprehensive application with the Florida Department of Environmental Regulation (DER), which serves as the coordinator of the process. The application must describe in depth the environmental impact of the project. DER then contacts other affected state and local agencies, including the state public service commission, and gives them an opportunity to comment according to a preestablished schedule.

DER is responsible for processing many environmental permits under preexisting state law. Based on its own evaluation and comments from other bodies, DER then prepares a written analysis and a proposed permit, called "conditions for certification," that are subjected to a public hearing. All permitting issues are dealt with at the hearing, and any agency from which the project proponent seeks a variance or exemption must be involved.

The hearing officer makes findings and enters a recommended order that is forwarded to the governor and cabinet for final decision. Once signed by the governor, the certification operates as the single state license for building and operating the facility. Most important, the Development of Regional Impact (DRI) review process required for all major developments in Florida—a process often viewed by developers as cumbersome and delay-ridden—is bypassed.

Although no further state approvals are needed, the PPSA does not automatically replace local land-use controls. However, if local approval is not forthcoming, the applicant may appeal to the governor to override. This power has not been used.

The act, passed in the early 1970s after a Florida Power and Light plant was blocked, generally receives good reviews from everyone involved. Since 1973, 12 power plants have been built under the law's provisions. One of the act's architects believes that it has worked

> as a conflict resolution tool that forces everyone into one process. It doesn't preempt existing agencies but makes them communicate. Earlier bills tended to reorganize, merge, move employees. This one doesn't displace anyone. No one is fired, no new bosses. Industry benefits because if agencies aren't in it together they can impose contradictory and costly conditions. Environmentalists benefit because it tends to pull reviews together, making them more comprehensive.[21]

Utility executives, too, favor the law, although they are wary of the broad citizen-participation provisions that inevitably bring a project publicity. Environmentalists believe that reviews are generally better, but some complain that the process is very expensive because it is formal and requires legal representation. Others say it leads to trade-offs when three or four agencies in favor of a project gang up on a hesitant agency at a public hearing. They claim the recalcitrant office will often give in rather than be portrayed as a holdout.

The relative success of the PPSA provides an interesting contrast to a later, broader law patterned after it, the Florida Industrial Siting Act of 1979 (ISA).[22] The ISA is generally similar to the PPSA, but there are key differences. It applies to any major industrial project, not just power plants. Hearing and decision times are generally shorter under ISA, and there are provisions protecting permitted facilities from later changes in laws and regulations. ISA is not mandatory for all projects and preempts neither the existing DRI process nor local laws. To date, no industry has applied to use the ISA procedures, largely because it is viewed as another bureaucratic layer that will not ensure speedier decisions.

Washington

The state of Washington's siting procedure for major energy facilities is credited by supporters as the key to siting seven nuclear plants.[23]

A state siting council, consisting of a chairman appointed by the governor, 14 state agency heads, and representatives of local governments affected by the proposed project, is empowered to issue a facility certification in lieu of all other state and local permits.

Final approval authority rests with the governor. Affected agencies and governments have a voice in the process, but not veto power. And although federal permits for these major facilities must be sought independently, the council's enabling act requires it to coordinate activities with federal agencies. To that end, the council has adopted federal environmental impact statements as its own and has held joint hearings with federal agencies.

The Washington act is not without its detractors, however.[24] Some say that review quality has been sacrificed on the altar of speed and that the system is really just a way to circumvent regulatory requirements. Others allege that agencies asked by the council to comment on projects fail to devote as much time and resources to the process as when they are individually responsible for projects. Interestingly, the council recently rejected the so-called Northern Tier pipeline that is slated to cross the state, despite federal government claims of national need.

The Florida and Washington experiences demonstrate that siting acts can be made to work (at least for states), but not by ignoring laws and agencies. Decision-making powers are vested in a single entity, but agencies and governments that might be affected have a formal voice in the review process. In neither are environmental reviews discarded. In fact, in Florida, the process produces a more comprehensive analysis than did the previous approach.

No state, however, has yet established a successful siting process for all major industrial projects. The Florida and Washington laws apply only to power plants, facilities generally alike in their design and built by companies already highly regulated and familiar with government intervention into their affairs. Such projects differ from many other industrial developments in that the public and regulators can usually determine more readily whether they are needed and offer desired benefits (although in Washington the state appears to have authorized too many plants).

Can Truncation Work for Industrial Siting?

Attempts at truncation continue to attract support, and at first glance, some of these "impatient" reforms make sense. For example, they might have helped clean up the overlap and lack of coordination among state and regional agencies and the local government in the Dow case. And, as that case shows, local governments generally cannot adequately evaluate the environmental and land-use impacts of major industrial projects.

Cutting back on citizen participation is likely to cause more environmental problems in the final analysis, however. Governmental reforms to make sure decision-makers consider all sides of an issue were developed in response to the resentment and open protests of the 1950s and 1960s by citizens who were allowed to say little about major projects—be they highways, high rises, or high-voltage transmission lines. Although hearing procedures may need improvement, perhaps by making them less adjudicative and by holding them early in the siting process, a return to a closed decision-making process would be ill advised.

Experience with the Minnesota Siting Act is instructive. Farmers felt their views were being ignored in the siting of an extra high-voltage power line across the state from a power station in North Dakota. In retaliation, they vandalized utility equipment along the proposed right-of-way.[25]

The European experience also teaches that stifling citizen participation will eventually backfire. One international survey of environmental conflicts found that

> violence has been three times, and mass demonstrations four times, as frequent abroad as compared to the United States. The tactics of opponents in Western Europe, particularly of antinuclear activists, have often been quite flamboyant and/or violent—extended plant site occupations in France, West Germany, and Switzerland; bombings and sabotage of plants, construction equipment, and high tension lines; terrorist attacks on executives of polluting corporations and officials of lax government agencies; and massive demonstrations, at times involving up to 50,000 protesters, led by a belligerent new breed of roving European "ecology troopers."[26]

The report concluded that "these kinds of tactics perhaps reflect high levels of frustration resulting from exclusionary political and corporate decision making processes, as well as the relative absence of means to redress grievances effectively."[27] The Barlöcher chemical plant siting dispute is a perfect example. Shut out of the decision-making process within the French central government, local citizens who opposed the project took to the streets and occupied the construction site for several months. Their reaction finally persuaded the central government to kill the project.

Similarly, there are serious shortcomings in proposals to strip state agencies or local governments of decision-making power. Federal agencies, as well as state and local governments have caused delays in some siting disputes. In addition, state or local governments excluded from the environmental decision-making process can usually find other ways to slow or stop a project. As the European case studies discussed in chapter 4 illustrate, local governments can thwart projects even though the central or state governments have their hands on the formal levers of power.

In this country, when the tide appeared to favor passage of legislation creating an EMB, several states were already planning how to "EMB-proof" state laws to prevent federal override of a state agency that had not met a decision schedule. If the EMB had started to intervene, the state agency might simply have quickly denied the permit application.

Moreover, many in industry viewed the powers of an EMB skeptically, realizing the implications of operating in a state or locality whose objections had been overridden. As one oil company representative put it, designation of a project for priority treatment would have been the kiss of death, especially in the West.

Another reason it is impractical to get state and local governments or particular agencies out of the permitting process is that the federal government is not generally equipped to evaluate energy- or industrial-facility siting proposals alone. Federal agencies are less likely to be aware of problems peculiar to a site, be it water availability or secondary-growth impacts. For example, an EMB staff in Washington, D.C., could never be expected to apply local land-use laws for a synfuels plant in Colorado or an oil refinery in South Carolina. Again, there is much to be said for the point that those who will have to live with the impacts of a big project should have a real say in the permitting process.[28]

Proposals for state siting councils raise many of the same questions that federal truncation does, including serious legal issues, particularly where local governments have home-rule authority or their powers are derived directly from state constitutional authority.

Industry officials have been skeptical of proposals for states to override local land-use regulations. By a large margin (75 percent), members of the Industrial Development Research Council, an association of industry-siting specialists, voiced opposition to the American Bar Association's state siting-agency proposal. As one facility planner said:

We don't want an umbrella agency making decisions that affect the local community. Once we get a plant built, we have to live with the people in the community, and if they're unhappy about a state-level decision, you can bet we're going to be unhappy, too. We'd like to be able to iron out our differences with the local people we're going to live with.[29]

Furthermore, businesspeople have expressed qualms that preemptive siting authorities might actually mire the permitting process in additional bureaucracy.[30] In Montana, some in industry have derisively called the Major Facility Siting Act the "Major Facility Prevention Act."[31] Experience in New York has been similar. Utilities helped enact a state power-plant siting act to solve the multiple-permit problem; now, they claim, it takes twice as long to get permits.[32] One major study of the energy-facility siting process notes, "Many in the energy industry feel that they have been betrayed, that siting legislation has increased the uncertainty, delay, and expense of constructing major facilities."[33]

Indeed, the track record of most efforts to truncate the permitting process is poor indeed. The tremendous opposition generated by current proposals should stand as a clear warning to would-be reformers. Ignoring citizens, public-interest groups, regulatory agencies, or environmental laws probably will not make the permitting process any more efficient, and, if it does approve some facilities that might otherwise have been turned down, it certainly will not make water effluents or air emissions from a mine or plant any less noxious to nearby residents.

SITE SELECTION BY GOVERNMENT AGENCIES

As an alternative to EMB-like super-siting councils or agencies, why not have government agencies identify and buy environmentally acceptable sites for chemical plants, mineral development, oil refineries, and other major projects? After all, advocates have reasoned, communities have been establishing industrial parks for years to attract industry, particularly light manufacturing; it would not be that big a step for government agencies to relieve industry of evaluating environmental impacts and at the same time maintain a bank of acceptable sites that could be drawn on at industry's convenience. Problems like the ones that arose in the HRECO and Dow cases, where the sites chosen appeared to be poor alternatives from an environmental perspective, might be avoided.

Although this idea has some appeal on paper, experiences in several American states and European countries that have tried such a scheme for major facilities reveal some serious difficulties.

Maryland

In 1971, Maryland adopted a Power Plant Siting Act giving the state Department of Natural Resources (DNR) the obligation to evaluate the environmental impact of proposed sites and to purchase four to eight acceptable sites around the state for future power-plant development.[34] The operation is financed by a small surcharge on utility bills. Once a site is purchased, the property can be used without regard to local zoning laws. The theory is that power companies will know with certainty a few sites where they can go and thus avoid uncertainty and delay.

Although the program has some strong points—all sides agree that it provides unbiased, scientific analysis of impacts—it has run into some serious difficulties. Initially, public hearings in the locality of the chosen site were not required, and no provisions were made in the law for disposition of a site purchased by the state but not used by industry. Amendments to the act in 1978 helped solve these problems, but the site selection and analysis process remains time-consuming and expensive.

The DNR, while it evaluates impacts and purchases sites, does not have overall permitting authority. Final approval must come from the state public-service commission in a separate regulatory process (into which the DNR analysis feeds). The required impact studies have also been expensive—$500,000 for a single site in western Maryland—and rising land values in the 1970s made site purchase expensive, especially when utilities were not required to buy the sites selected by the state. (In fact, one utility, despite opposition from environmentalists, expanded at a site it owned rather than use state land.) Some critics have also said that, since a state agency is involved in selecting sites, its choices are open to political pressures unrelated to either the legitimate siting needs of utilities or concerns over environmental quality.

When these problems are added to several other difficulties, ranging from opposition to specific sites by local governments and environmental groups to rapidly declining electrical demand projections, it is not surprising that only one site has been acquired and that even it stands vacant.

San Francisco

In the wake of the Dow dispute, the Association of Bay Area Governments in San Francisco toyed with a scheme similar to Maryland's that would have identified and graded all available industrial sites in the region.[35] An industry could have applied for a permit to build a new facility wherever it pleased, but it would have received bonus approval points by locating in areas with attributes such as high unemployment, good transportation, and absence of environmental problems. Political infighting killed the idea before it could even be officially proposed. Practically every local government wanted to be designated for growth, particularly if the industry involved was a clean one.

Europe

The experience of the United Kingdom with industrial location controls is also instructive.[36] The country has for many years imposed strict controls on industrial location and offered economic incentives in an attempt to lessen congestion around London and to revitalize depressed areas of the nation. Evidence shows that the controls have been effective in bringing about substantial redistribution in industrial growth.[37]

However, the movement of industry from London has contributed to the erosion of that city's economic base. In addition, industry claims not only that some firms have been forced to build in sites where they are at a considerable competitive disadvantage but also that the controls have discriminated against companies that cannot or do not want to move.[38] Recent economic problems in the country have led to a relaxation of the controls.

There are some positive attributes of close government involvement in selecting sites for major facilities, notably good environmental reviews. Overall, however, experiences in the United States and elsewhere do not argue strongly for governments selecting sites for new facilities. Perhaps the fundamental question is whether federal or state agencies are knowledgeable enough to select sites for industrial growth. As one chemical industry representative observed: "I've been building plants for 25 years and it gets tougher all the time. I'll be happy to give that job to government if you can persuade me they have the experience and can do a better job."[39]

HOG-TYING THE REGULATORS

If streamlining or truncating the process does not work, and giving government site-selection power fails, what then? Another popular approach to regulatory reform is to make it difficult, through a variety of methods, for regulators to regulate. The theory seems to be that, if overzealous regulators are brought to heel, then industry will have fewer problems in getting plants built.

Attempts to restrain regulators are not new. Congress passed the Administrative Procedure Act in 1946 to counter what it saw as unchecked discretion of regulatory agencies. Current attempts to curb environmental regulators take several forms. Some would tie up the regulators by enacting complicated procedures and rules, such as cost-benefit analysis requirements. Others would remove the presumption of validity normally attached to regulators' actions or cut funds and personnel so that regulators simply lacked the time or resources to overregulate. Will such efforts succeed? Will they get at the real problems?

The most well known example of regulating the regulators is the proposed Regulatory Reform Act, introduced in 1981 by Senator Paul Laxalt of Nevada and passed by the Senate in 1982.[40] In a nutshell, Laxalt's proposal, if passed by the full Congress, would require:

- Strict cost-benefit analyses of all major regulations according to rules set by the Office of Management and Budget. If benefits could not be proved to exceed costs, the regulation could not be enacted. Parties could challenge in court the cost-benefit techniques utilized.
- "Sunsetting" of agencies or programs so they would automatically expire unless saved by Congress or unless they could prove the benefits they brought about outweighed the costs.
- Congressional review of regulations, with the right to veto any regulation before it took effect.
- Broader court review of regulatory actions.

The Reagan Administration has already taken similar steps for executive-branch agencies. Executive Order 12044, signed by President Reagan in February 1981, directs all federal agencies to prepare a cost-benefit analysis for each major rule that they propose.[41] Then, to the extent permitted by law, the agencies must choose the most cost-efficient alternatives. The Office of Management and Budget oversees implementation of the order and must "sign off" on all regulations.

Another approach, which differs markedly from the first, is found in the so-called Bumpers amendment, named for its sponsor, Senator Dale Bumpers of Arkansas.[42] In contrast to Senator Laxalt's bill, the Bumpers amendment focuses on the tail end of the regulatory process by removing the presumption of validity that courts normally attach to actions by regulatory agencies. Thus, if someone judicially challenged an environmental regulation promulgated by the EPA, under the proposed amendment the regulation would fall unless the court could be persuaded there was substantial factual support for it. No longer would courts defer to administrative actions in the absence of evidence that they were unreasonable, arbitrary, or capricious.

Senator Bumpers claims that both business and environmental groups should support his initiative. "It's not the only approach to regulatory reform, but it is the one most likely to provide the kind of relief the business community wants and is entitled to."[43] Further, he says, it is the answer to environmentalists' prayers because under the amendment they could do something about their long-standing criticism that the EPA fails to issue tough regulations to implement existing law.

Will these initiatives work? Will they solve the problems in the environmental and land-use regulatory process that are holding up industrial development and preventing effective impact assessment? Almost certainly not. Rather than addressing the real problems, these proposals would make the regulatory process so cumbersome that regulations would never get off the starting blocks. As one critic quipped, they look more like lawyers'-relief acts than serious attempts at reform.

Ironically, these proposals would probably be detrimental to *both* industry and the environment. Consider the possibilities if, as required by the Bumpers amendment, burden of proof were switched to agencies: A federal agency, after extensive hearings, would then produce an environmental impact statement and, based on it, approve a major industrial project. Environmentalists might then challenge the action, arguing that the decision was arbitrary and the impact statement inadequate. Under the Bumpers amendment, the court would be required to review *in detail* all the facts behind the agency decision, including the environmental impact statement and the transcripts of all hearings. The court might also rehear testimony so it could make an independent judgment based on everything in the record. Result: the suit would languish in court, perhaps for several years. Then, in some cases, the agency decision would be reversed, and the case remanded to the agency for additional fact-finding.

Of course, the plot could work the other way so that industry could challenge denial of permits. Yet the remedy would not likely be approval of a project, but simply further hearings and investigation. Benjamin Civiletti, attorney general in the Carter Administration, summed up the effects:

> This would directly undermine the value of creating agencies as specialized, expert decision-making bodies. It would throw to the courts for reexamination the volumes of evidence on highly technical factual and policy issues that are generated in the most fundamental rule-making. It would raise incessant uncertainty about the law, fostering more and more complex litigation challenging agency action.[44]

The strict cost-benefit approach required by the Laxalt legislation and the OMB might stop some federal environmental regulations from going into effect, perhaps to the delight of a particular industry. Of course, nothing in the Laxalt bill or OMB procedures could keep a skeptical state or local government from refusing an industrial project on other than environmental grounds.

But just as the Laxalt bill would make it difficult to promulgate new regulations, so would it stop attempts to improve existing ones. Thus, the EPA's celebrated "bubble" policy reform, designed to reduce the compliance burden on industries with air-pollution problems,[45] might take years to implement: the Laxalt legislation would allow environmentalists to challenge the cost-benefit equation, assuming it was positive, and the Bumpers amendment would remove the presumption of validity that now accompanies all agency acts. In fact, promising regulatory-reform initiatives might be stopped in their tracks without specific congressional approval.

This is not to say that cost-benefit analysis cannot be a useful tool in evaluating proposed regulation. But Laxalt's proposal makes cost-benefit analysis an inflexible rule instead of a helpful tool, just as the Bumpers amendment grasps for improvement by simply shifting the burden of proof rather than more closely analyzing needed improvements in administrative procedures.

Perhaps the Laxalt and Bumpers proposals should come as no surprise. Sweeping reform measures like these are more alluring than incremental, less flashy changes, even when the low-profile reforms offer far more promise of improvement in the long run. Fine-tuning just does not win many votes or attract much media attention. Furthermore, in a Congress filled with lawyers who have constructed a complex regulatory system that is adjudicatory and adversarial in nature, "reforms" might be expected to take a similar path. As one critic observed:

> [The Laxalt bill] bears a strong resemblance to the techniques of regulation itself. These measures would specify in great detail exactly what procedures each agency would have to follow in its rulemaking, what subject matter it would have to consider, and what statements it would have to include in official pronouncements. . . . There is no reason to believe that they will be any more useful as a means of controlling government than they have been as a means of controlling business.[46]

Again, this approach to regulatory reform fails to differentiate among agencies and their missions, or the real problems each faces in the regulatory process. Congress appears frustrated in its attempts to improve existing laws substantively and thus turns to what look like procedural changes that promise to accomplish a similar result. But as Senator Thomas Eagleton of Missouri cautioned in testifying on the proposed Regulatory Procedures Act of 1981 (the House equivalent of the Senate's Regulatory Reform Act):

> The first temptation is to use neutral-seeming procedural changes as vehicles for far-reaching change in particular substantive statutes. . . . Perhaps it can be demonstrated that there should be a cost/benefit standard in the Clean Air Act. However, that proposal should be considered extensively and in that specific context by Congress. . . . To consider the matter in a less focused context would risk imposing an unwise result on the specific substantive law and other essential regulatory decisions not the primary target of cost/benefit ratio proponents.[47]

In a parallel situation, a nuclear regulatory commissioner observed that the Reagan Administration's attempts to speed the siting of nuclear power plants by cutting back on regulatory oversight are based on the "cockeyed" premise that "the regulators are what is at fault, and what is needed is to stop the regulators. . . . Nothing is going to be so detrimental to nuclear power as the public deciding they can't rely on the regulators. There are enough people who believe that already."[48]

The same can be said for regulations governing industrial-plant siting generally. The Administration and Congress would be better advised to exercise oversight responsibilities with less haste and greater focus. This advice is easy to offer, but it will require that environmentalists and industry search together for specific reforms they can agree upon. That will be difficult given the increasingly polarized debate over environmental laws.

PUTTING IT ALL TOGETHER: A STRATEGY OF COOPERATION

Impatient reforms and cure-alls have fallen short of their often lofty goals. In the process, they have created as many problems as they have solved, offering false promises of quick relief. Problems in the regulatory system have persisted and led to experiments with less drastic forms of environmental regulatory reform that rely more on cooperation and compromise.

The experiments with "quiet," incremental reforms indicate that governments and industries can clean up their own backyards and alter the regulatory landscape in a significant, positive way by working together to forge a rational siting process that includes all major parties in the decision-making.

How the Colorado Joint Review Process Was Conceived[49]

The Colorado Joint Review Process (CJRP) was one of the first attempts at quiet reform by industry and government. The techniques devised for the CJRP offer important guidance to a new generation of reformers.

Ironically, the CJRP had its origins in a bitter dispute not over an industrial facility but over a major ski resort, at Beaver Creek in northwestern Colorado.[50] Planning for Beaver Creek started in the early 1970s and was accelerated in 1972 when the site was selected for the 1976 Winter Olympics. But Coloradans, led by a state legislator and college professor named Richard Lamm, rejected the games in a November 1972 vote. Despite that setback, environmental impact studies for the resort continued. Everything went smoothly in 1973; in 1974 the U.S. Forest Service issued a final environmental impact statement, and state agencies gave the project a green light. But it was not that simple.

Lamm rode into the governor's mansion in November 1974 on a wave of concern about land-use and environmental degradation in the state. One of his first acts was to challenge the federal environmental impact statement and state approvals for Beaver Creek development. Environmentalists threw support behind him.

However, instead of fighting the opposition, Vail Associates— the project developer—embarked on a strategy of cooperation. One company official explained:

> Quite simply, rather than fighting or habitually refuting the environmentalists' concerns, we made a concerted effort to demonstrate either the negative impacts they predicted would not—*could* not—happen, or we came up with mitigation that corrected anything they felt would go wrong. Instead of bitching back and forth, letting our emotions guide us, we tried to answer their concerns reasonably. To do that, we had to have the facts.

Vail Associates thus undertook a new comprehensive environmental impact study, all the while keeping informal lines of communication open with state agencies and environmentalists. By 1979, the resort was open, and despite some minor quibbles, Lamm and conservation groups were generally pleased with the outcome.

At about the time the Beaver Creek project was surmounting its last hurdles in the late 1970s, Colorado was coming under intense pressure from mineral and energy development. Lamm was also being raked in the Republican state legislature as an anti-growth environmentalist whose state agencies were tying companies' development plans up in knots. This is when the CJRP idea was born.

AMAX Tests the CJRP

Governor Lamm and Harris Sherman, then head of the Colorado Department of Natural Resources (DNR), which had permitting authority over most mining- and energy-facility proposals, were casting about for ways to deflect anti-growth criticism. They knew that overlapping and contradictory regulations were causing problems. Sherman also realized that cooperation between state agencies was virtually nonexistent and that this was slowing projects down. But Lamm and Sherman felt that companies had a responsibility to bring the public and government agencies into the project-planning process earlier, before plans were set and opinions polarized. They wanted to change the way corporations thought about big projects.

According to one DNR representative, "They [corporations] are often like a big kid with big feet when it comes to the regulatory process." The Beaver Creek experience offered some valuable lessons, and perhaps an alternative approach, but Lamm and Sherman felt there were ways to do it better. All that was needed was a firm willing to cooperate and test their new ideas. That's when AMAX, Inc., a multimillion-dollar minerals and mining company, entered the picture.

In January 1978, AMAX notified the state that it was evaluating, and would probably develop, a rich discovery of molybdenum (a "space-age" mineral used, among other things, to strengthen steel) in Gunnison County near the small, southern Colorado town of Crested Butte. Local residents immediately protested, vowing to stop the proposed mine and mill, which would bring over 1,500 workers to the area by 1990 and cause the county's population to increase from 12,000 to over 18,000.

AMAX's reaction was not to bare its teeth; instead, it decided to replicate a highly acclaimed planning effort it had used in developing another molybdenum mine near Henderson, Colorado.[52] That effort was based on open planning and thorough environmental impact analysis and mitigation. A vice-president of Climax Molybdenum (an AMAX subsidiary) later recounted why the company decided to play guinea pig:

> Although we were confident [based on AMAX's Henderson experience], we were also fully aware that during the past seven years more environmental and land-use regulation had been passed in this country than in the previous 100 years. This meant that many more agencies would be involved in the review of the Mount Emmons Project than for the Henderson Mine, which had just been completed after a 10-year development period. We also knew that the success of agencies and project proponents in dealing with these new laws had been uneven and that both government and industry representatives were becoming increasingly frustrated in their efforts to administer and comply with these laws. Although we felt we could satisfy most permit requirements, we had a nagging feeling that some agencies with new regulatory authority were tending to view their responsibilities more expansively than was intended, and that various local, state and federal agencies did not understand the mission of other agencies which had environmental protection responsibilities. There was a tendency for several agencies to want the last bite at the apple especially if they viewed themselves as the agency best equipped to assess impacts and regulate in the environmental protection area. Agencies were not well coordinated in their review of some projects and this caused some agencies to be very critical of other agencies in their handling of permit requirements. This situation was recognized and several agencies, including the Colorado Department of Natural Resources [DNR], were actively exploring ways to improve cooperation and coordination. The atmosphere was right in late 1977 for agencies and project proponents to find better ways for coordinating the planning and review of major resource development projects.[53]

This is not to say there weren't some skeptics within the company:

> Arriving at an open planning approach was not as easy as I may have made it sound.... Within my company we wondered if encouraging early government and public involvement would do little more than educate opponents sooner.... Under some circumstances, that argument might have merit,

but we decided we were capable of managing an open planning process and would have less credibility if we were not always prepared to discuss the project, answer questions, and encourage input from agencies and the public. Our assumption was correct.[54]

Thus, the thinking of state and company officials seemed to be in tune, and when Lamm and Sherman invited AMAX to participate in a prototype program to improve and expedite the process of developing mines and other big projects in the state, the corporation agreed.

The keys to the process, which was set in motion by mid-1978, were openness, informality, and voluntary participation. No new laws would be enacted or existing laws waived; no big bureaucracies created; no party forced to participate. Harris Sherman described it as "simply a management system designed to increase the efficiency of government decision-making."[55] The company and agencies would put their cards on the table early to avoid delays later.

A small coordinating office within DNR was established to oversee the effort and to act as coordinator. One of the first things the CJRP office did was to negotiate with other levels of government to establish lead agencies. DNR would lead for the state in this case, the U.S. Forest Service for the federal government, and Gunnison County at the local level. The county would act as overall lead agency for the four-member review team (the three agencies and AMAX).

When lead agencies were assigned, the three signed an agreement pledging to cooperate, communicate, and coordinate. AMAX was not asked to be a signatory (an omission that would be changed when other companies chose to use the CJRP in the future). The CJRP office was then assigned to monitor the process, to see if each party was holding up its end of the deal, and to act as neutral convenor for the CJRP team and public meetings.

After this organizational phase, AMAX began holding separate informational meetings with the project team, other agencies, and the public. (In later reviews, there was at least one consolidated meeting.) At these meetings, AMAX described the project in some detail—the mine, a processing facility, and attendant developments, including a large power line and a mill-tailings pond. The public was also given a chance to ask questions and voice concerns about the project and its impacts.

Using the agency and public input, AMAX redesigned elements of the project—relocating the mill and changing access corridors—to reduce environmental impacts. The company also learned how strongly local citizens and governments felt about dealing with secondary-growth impacts and maintaining the quality of life in their town.

Because of these strong feelings, AMAX agreed to pay for an independent consultant to analyze socioeconomic aspects of the project for the county. This was a real change in the way AMAX and other companies usually do business, but it gave the public confidence in the results and saved AMAX from first doing an in-house study and then paying for another later when the public demanded a neutral analysis.

AMAX also began taking steps to deal with secondary growth. It purchased an 800-acre parcel of land within Gunnison's city limits to ease housing problems and to guard against sprawling development that would overload city services and deface the landscape. It also optioned a 300-acre ranch outside the city limits that could be developed to house 1,200 units, including some mobile homes.

About one year into the process—in mid-1979—AMAX signed a statement of responsibilities with various agencies at all three governmental levels. The aim was to develop a common data base and make clear who was to do what, thus giving AMAX some certainty in its project planning. In return, AMAX agreed to continue its open planning approach. Project planning and work on the federal environmental impact statement continued while AMAX began filing for necessary air- and water-quality permits.

The next step in the CJRP was to prepare a decision schedule to coordinate AMAX's proposed timetable for further planning and development with the regulatory process. However, by this time, it was becoming clear that the CJRP needed some revamping if it was to work for other projects. The statement of responsibilities should have been signed earlier to reduce confusion. The team and public meetings could have been more effective by separating the descriptive stage, in which company officials explained the project to the public and agencies, from the one involving "scoping," where environmental and socioeconomic issues were identified.

There were other pressures for change besides those unearthed by AMAX's trial and error. Several other companies were lining up to file applications for huge projects as Congress appeared ready to spend billions to create a new synfuels industry based largely in Colorado and Wyoming. Moreover, President Carter had proposed his EMB, which Colorado officials feared might run roughshod over state permitting procedures, and the legislation seemed sure to pass. Consequently, the state took steps to revamp the CJRP process in midstream.

In June 1979, the architects of the CJRP began dissecting the process with help from an experienced advisory group. In early 1980, several state legislators proposed an eventual consolidation of state agencies and creation of a state siting council for energy projects, which helped spur the remodeling.

By April 1980, with financial assistance from the U.S. Department of Energy, state officials had developed a model process (see appendix B) that covered four types of development: a coal mine, commercial oil shale, a uranium mine and mill, and a metal mine and mill. Model decision schedules were also charted.

AMAX was the first to buy into the revamped process even though it was well along in project planning. Since AMAX had completed several steps in the preexisting JRP, the first action was to establish an informal decision schedule for the project. This was done in mid-1980.

Rating the CJRP

How have AMAX and the CJRP fared since the 1980 model process was developed? How do participants in the process view its record?

AMAX's planning proceeded relatively smoothly throughout 1980 into 1981. Gunnison County appeared satisfied with mitigation measures AMAX proposed, and the state Department of Public Health issued air-pollution permits in February 1981. The U.S. Forest Service issued a final impact statement in the fall of 1982 that was generally favorable towards the project, and reviews by other state and federal permitting agencies commenced.

The town of Crested Butte—a small high-country hamlet surrounded by mountains of uncommon splendor and unlimited recreational opportunities only a few miles from the mine—did cause AMAX some problems. It passed a watershed-protection ordinance requiring a permit from the town for activities that might affect the town's water supply.

AMAX had studied possible alternative sources and offered to help the town relocate its water supply even before Crested Butte acted. As a result, a Colorado trial court issued a preliminary injunction staying the ordinance in June 1981, allowing AMAX to continue preliminary site work and planning. Although that case was still pending before the Colorado Supreme Court in mid-1983, AMAX officials say they are confident of eventual success on the project's merits.

Despite ups and downs in the process, company officials say the CJRP is a good idea. An AMAX vice-president who helped conceive the company's progressive project-planning approach said at a 1981 press conference: "This is the one place in the country which has developed a new regulatory mechanism that really works."

He and other corporate representatives like to compare the CJRP to AMAX's experience in Minnesota, where they say a proposed copper-nickel mine was delayed for two years in the permitting process. "We think this [the CJRP] has more promise than anything we've seen," said another AMAX vice-president. "The usual practice is to go from agency to agency, bouncing back and forth from state, local and federal offices. You end up chasing your tail around in a circle."[56]

But just how much time was saved? Some state officials predicted the permitting process could be cut nearly in half, perhaps saving two years overall. That prediction looks optimistic now, in part because the CJPR was new and some steps were out of synch in AMAX's case. In fact, AMAX took two years to reach the point in the CJRP process that the new model manual says should take nine months. But company officials still feel the draft environmental impact statement was probably completed four to six months earlier than it would have been without the CJRP.

In addition, officials believe ultimate community acceptance is more important than speed and that the CJRP has given the project high visibility and therefore priority within government agencies.[57] They say that, even if permits are eventually denied, the company will be in a better position to solve problems with agencies or opponents more quickly, either by negotiation or in court, because of the positive public record that has been established.

The process has been costly for AMAX, but because of the high value of the minerals involved (perhaps over $4 billion), the company has spared little expense to win public support as well as official approvals. It is not clear that other firms will be willing or able to go to the expense AMAX has, although reports from later CJRP projects are encouraging. Clearly, the CJRP is not designed for small or even medium-sized projects.

Although top AMAX officials and the Mount Emmons project manager are solidly behind the CJRP, there has been some grumbling within the company. The project manager said that "people back East [at the company headquarters] may think things have gone a little haywire out here," so there is a constant public-relations battle to show the project is proceeding. There have also been reports that middle-level employees have criticized the process to their counterparts in other firms. The project manager attributes this griping to the fact that "some of our people in Crested Butte just can't believe how much things have changed since they opened mines in the 1960s."[58] Overall, however, AMAX likes the process and would probably do it again, although company officials say the worst thing to do is to oversell it as a panacea for all permitting problems.

What of government agencies? Gunnison County officials almost uniformly praise the process, although some are seeking changes in the project's size and other details. They praise AMAX's open approach and say the CJRP is a step in the right direction. "They tell us what they are planning, and we avoid government by ambush—we let them know what we're doing."[59] Of equal importance, they stress, is having local governments take the lead in the process: "This is where it will happen. Not in the state capitol or in Washington, D.C. If the EMB came in here and said this was the right site, we'd throw them out."[60]

AMAX's attention to secondary-growth impacts has also won it kudos. Even town officials from Crested Butte who adamantly oppose the project have good words for the process: "CJRP is good for the public. It's an open approach. And if a place has to be mined I'd rather have AMAX doing it, but we are making them do a lot of things I don't think they would have done on their own."[61]

The head of the U.S. Forest Service's environmental-impact-assessment effort likes the process because, he says, it is open and not a chess game.[62] He also feels that the CJRP has instilled discipline in the agencies to act expeditiously and has resulted in better environmental impact analysis and mitigation efforts.

Like the other players, however, Forest Service personnel have run into some problems and see ways to improve the process. According to one official, some confusion among federal and state agencies early in the process apparently slowed things down. Earlier warning that the process was to be used would have helped the service gear up faster.

Many people say the federal side of the CJRP has worked because the Forest Service has been very conscientious and has maintained an image of independence and objectivity in the public eye. Some question whether things would have worked as well if the Bureau of Land Management had been the lead federal agency because, they say, it has a reputation for being very pro-mineral development.

State officials, particularly in the governor's office and in the DNR, are high on the CJRP. One of the major benefits they perceive is better quality project reviews. "The trade-off involved is more certainty for better environmental assessment and more public participation," says the first director of the CJRP office in DNR,[63] who adds that another benefit is more efficient use of time: "It has substantially improved the coordination among agencies. It has also put agencies in the limelight so that if they botch things up they will be scrutinized and held responsible." He disagrees with some environmentalists who suspect that the process is too development-oriented because of the pressure it puts on agencies: "It doesn't guarantee a yes, only that decisions will be made on time. The process may have a schedule, but that helps environmentalists too by laying out a roadmap so they can gear up for important points in the process."

The former CJRP director concedes, however, that the process is costly and that the agencies are still learning the ropes. "When the JRP is in full swing it could use five people, but the funding question is open because the state legislature won't enact anything." Because of the time and money involved, he says that the state will be very selective in inviting major projects to utilize the CJRP in the future. (Like other state agencies around the country, the CJRP office itself has experienced a significant turnover rate. Both the director and the assistant office director recently left for positions in the private sector.)

Environmentalists in Colorado and with national groups have generally praised the CJRP as a preferable alternative to an EMB with power to override environmental laws. One notes: "The CJRP is noteworthy for what it *doesn't* do. It doesn't create a new bureaucracy, doesn't override existing laws, doesn't reduce public access to the decision making process."[64] However, some do complain that an expedited schedule and fewer hearings might make it more difficult to stop projects they strongly oppose and that the process may promote too cozy a relationship between the regulated and regulators. Some citizens have also argued that the large number of hearings and meetings that are a feature of the process are time-consuming and leave them less chance to analyze the project.

Overall, the CJRP gets better than passing grades from most of the participants in its first run. It does have much to offer to both industry and environmentalists, but trade-offs are involved. The process is costly for state and federal regulators (as well as industry), and it hasn't insulated the AMAX project from legal challenge.

In fact, AMAX has not begun serious mining yet; a softening world molybdenum market—AMAX's molybdenum prices dropped from $9.50 per pound when the company first proposed the mine to about $5 now—forced AMAX to delay construction for two years, until 1984 at the earliest. Moreover, although the Forest Service issued a final environmental impact statement generally favorable to the project, several permits and approvals are still outstanding.

Yet to proponents of several other major projects, the CJRP has looked more promising than business as usual. A new class of project hopefuls—including energy developments, ski resorts, a nahcolite mine, and a low-level radioactive waste disposal facility—has started to proceed through the process (although the energy projects have been slowed or shelved recently due to a slack market).

An official from Chevron, which has proposed an oil shale facility in western Colorado, listed the pros and cons of the CJRP from his firm's point of view, explaining that on balance the advantages outweighed the disadvantages.

- Although the CJRP provides for additional public participation opportunities that may enhance the effectiveness of opposition challenges, the company concluded that the risk of successful legal challenge and delay is substantially diminished, and the early discovery of issues of problems of key public concern provides the company with adequate time to address those concerns, thereby minimizing delays.
- Although added procedural requirements are required, these new procedures minimize the possibility of "government by ambush" and, thus, delay. A cooperative attitude between the company and agencies is preferable. Such an atmosphere will enhance communication and the productivity of negotiations on tough issues that will come later in the process.
- Even though the process is voluntary, agency participation and commitment has been exceptional. Agencies, like the company, prefer the CJRP approach to the traditional adversary approach. That atmosphere is more productive leading to more acceptable results. The better a company communicates, the stronger the agency's commitment becomes.
- Early and continuous contact with government agencies, which could lead to attempts to force early design changes or fear by agencies of

prematurely committing to the project, has in practice helped minimize delay and ensured sound, defendable decisions.
- The project decision schedule may give advance notice to opponents and is unenforceable, but it is an excellent planning tool that fully discloses all aspects of the permitting process well in advance, thereby providing a record that more than adequate notice was given. It diminishes the possibiity of procedural violations because it requires an accurate reflection of each agency's legal process. Although unenforceable, the schedule creates momentum to complete the process as scheduled, assuming no major problems arise.[65]

Just as other companies have expressed interest in the CJRP, so have other states. But there are important questions about the process that other states must grapple with in adapting this tool to their needs: Will a voluntary CJRP process work for industries that, unlike mines and synfuel plants, are "footloose" and can locate elsewhere? The projects currently going through the CJRP process depend on site-specific natural resources and therefore have less ability to negotiate or extract concessions from governments. What will the CJRP do for smaller projects? It is probably too expensive to be used often. What if a particular agency, say a local government with authority over crucial permits, refuses to participate? And what if laws change in midstream or agencies fail to follow a decision schedule? States considering the establishment of their own joint review processes must carefully analyze such issues if new systems are to flourish.

Perhaps the biggest stumbling block to wide and effective use of CJRP-type reforms is that they will be oversold. As one business professor has observed:

> The bulk of most speeches, press releases and articles on the CJRP process have it as very close to nirvana or utopia. . . . It is not the miracle drug or panacea we have all been waiting for. . . . It may raise the probability of achieving [solutions to siting problems], but it certainly doesn't make them sure bets. . . . Don't get me wrong. Raising the probability of achieving more efficient and effective permitting, environmental assessment, citizen participation and conflict resolution is tremendously exciting and important.[66]

This evaluation seems correct. Probably the worst thing that can be done is to take a "quiet" reform like the CJRP and ballyhoo it as something more revolutionary, a tag that could help eventually to put it in a box with other discarded reform efforts.

CONCLUSION

Two types of responses to the new realities of plant siting have been tried: cure-alls, characterized by simplistic or impatient thinking; and quiet reforms, highlighted by cooperation between regulators and the regulated.

The cure-all reforms are alluring, but they have fallen short for a variety of reasons. They illustrate two important points that would-be reformers should keep in mind.

- *Be wary of changes that will be perceived as weakening existing environmental reviews and safeguards.* Rather than facilitating thorough environmental reviews that would reassure the public, the cure-alls often cut reviews short, thereby giving the appearance that there may be something to hide or that neither the public safety is assured nor resources protected.
- *Avoid the lured of truncating the process.* Reforms that override governments, ignore laws, and exclude citizens will not work well except in a few special cases. Such efforts fail to recognize that established laws and procedures represent strongly held views and values that cannot be changed by running roughshod over them.

Beyond these, a host of less obvious points seem clear:

- *Be cautious of changes that might provoke entrenched bureaucracies and their clienteles.* Many state and local agencies were just learning the environmental ropes when they were hit with calls for major reforms that would shift responsibilities, destroy old bureaucracies and create new ones, and fundamentally alter established jurisdictional lines. Not surprisingly, they and their constituencies resisted.
- *Take care that reforms are aimed at the right problems and not focused too narrowly.* Many of the early reforms were put in place with great expectations, but later suffered when results were modest. Most focused narrowly on only one or two critical problems in the siting process, usually those attributable to government agencies. When success did not come quickly, the reforms lost their allure.
- *Watch out for interests that might try to torpedo reforms because they like the existing system despite (or as a result of) its shortcomings.* Many companies have ignored reform measures that are not mandatory (for example, the use of consolidated hearings) because they understand the existing system and see few

benefits from changing the rules of the game.
- *Recognize the limits of agency capabilities.* Super-siting agencies have often been given tremendous responsibilities without adequate resources. In other cases, notably the efforts to have agencies choose sites, the responsible officials simply have lacked the expertise to do so on a broad regional or statewide basis.

The strategy used by AMAX and the state of Colorado—though it is only a first-generation reform and certainly not the answer to all regulatory problems—shows that cooperation among government, industry, and other interests offers the most hope for successful reform. The CJRP has been successful because it incorporates some important ideas: better internal governmental management techniques like lead agencies, agency contacts, joint hearings, and a project escort/mediation service; early public participation; early corporate environmental assessment; and voluntary project decision schedules.

These techniques and others that can help in remodeling the 1980s permit process are explored in the next chapter.

REFERENCES

1. Over the past decade, much has been written about permit coordination, especially relating to residential and commercial developments. For example, the Urban Land Institute, under a federal grant, published Fred Bosselman, Duane Feurer, and Charles Siemon, *The Permit Explosion: Coordination of the Proliferation* (Washington, D.C.: Urban Land Institute, 1976). The Conservation Foundation followed with John H. Noble, John S. Banta, and John S. Rosenberg, eds., *Groping through the Maze: Foreign Experience Applied to the U.S. Problem of Coordinating Development Controls* (Washington, D.C.: The Conservation Foundation, 1977). Other useful references include Robert Healy, "Coordination: The Next Phase in Land Use Planning," *Journal of Soil and Water Conservation* (July/August 1976), pp. 140-45; and Nelson Rosenbaum, *Land Use and the Legislatures: The Politics of State Innovation* (Washington, D.C.: Urban Institute, 1976).

2. Camella Auger and Martin Zeller, *Siting Major Energy Facilities: A Process In Transition* (Boulder, Colo.: The Tosco Foundation, 1979).

3. Bosselman, Feuer, and Siemon, *The Permit Explosion.*

4. For an excellent review of the act, see James Jarrett and Jimmy Hicks, *Untangling the Permit Web: Washington's Environmental Procedures Act* (Lexington, Ky.: Council of State Governments, 1978).

5. The system is discussed in "Consolidated Permitting Regulations: Miracle or Mirage?," *Environmental Law Reporter*, vol. 10 (May 1980), p. 10092. It covers the Resource Conservation and Recovery Act's hazardous-waste disposal program, the Safe Drinking Water Act's underground-injection control program, the Clean Water Act's National Pollutant Discharge Elimination System, state-administered dredge-and-fill permits required by Section 404 of the Clean Water Act, and the

Clean Air Act's prevention of significant deterioration (PSD) program. For a humerous poetic analysis of the system, see William Anderson II, "The Consolidated Permit Two-Step," *Natural Resources Law Newsletter*, vol. 15, no. 1 (Fall 1982), p. 2.

6. Government White Paper, *The Reorganization of Central Government* (London: Her Majesty's Stationery Office, 1970); quoted in Noble, Banta, and Rosenberg, *Groping through the Maze*, p. 86, which discusses the creation of the department.

7. Barbara Alexander, *The Procedural Efficiency of Maine's Environmental Permit System* (Augusta, Maine: The Maine Land and Water Resources Council, 1978), pp. 115-16.

8. The description is based on interviews with J. Leonard Ledbetter, director of the Georgia Environmental Protection Division, and with Ronald Mayhew and Robert Rotan of the Georgia Department of Industry and Trade, Atlanta, Georgia, March 6, 1980. See also "One-Step Permits Woo New Industry," *Business Week* (September 11, 1978), p. 55.

9. Specifically, the Environmental Protection Division administers the following programs: air quality, safe drinking water, water control, groundwater allocation, surface-water allocation, solid-waste management, surface mining land reclamation, safe dams, erosion and sedimentation control, environmental radiation, and oil and gas deep drilling.

10. This information is from discussions with Aubra Anthony, deputy general counsel for programs and policies, the National Trust for Historic Preservation, and John Fowler, general counsel, the Advisory Council on Historic Preservation, in Washington, D.C., various dates, 1982.

11. In proposed amendments to Federal Land Planning and Management Act, the Bureau of Land Management would cut public review of land-use plans and make proposed planning criteria available on request, rather than publishing them for comment. The Department of the Interior has also proposed scaling back citizen participation under the Surface Mining Control and Reclamation Act. *Washington Post*, January 14, 1983.

12. American Bar Association, *Development and the Environment: Legal Reforms to Facilitate Industrial Site Selections*, Final Report by the Special Committee on Environmental Law (Chicago, Ill.: American Bar Association, 1974). The highlights of the report were discussed by the committee's chairman in J. Thomas Green, "Reforming Procedures for Industrial Siting," *American Bar Association Journal* (April 1975), p. 449.

13. American Petroleum Institute, *Energy Facility Planning and Siting in the Coastal Zone* (Washington, D.C.: American Petroleum Institute, 1979), p. 50.

14. See, for example, Michigan's Hazardous Waste Management Act, H.B. 4380 (January 1, 1980), which establishes a statewide siting board with power to preempt local zoning restrictions.

15. For an excellent discussion of these proposals and later versions of the legislation, see Ross Cheit, "Energy Law and the Environment," *Ecology Law Quarterly*, vol. 8 (1980), p. 727.

16. Testimony of William Drayton, Jr., assistant administrator of the Environmental Protection Agency, before U.S. House, Committee on Interstate and

Foreign Commerce, Subcommittee on Energy and Power, *Priority Energy Project Act of 1979: Hearings on H.R. 4573, 4499, and 4862,* 96th Cong., 1st sess. (July 20, 1979).

17. See Cheit, "The Energy Mobilization Board."

18. For a good summary of Reagan Administration truncation moves, see Felicity Barringer, "U.S. Preemption: Muscling in on the States," *Washington Post,* October 25, 1982.

19. For the saga of this proposal, see *Coastal Zone Management,* October 28, 1981, and February 3, 1982.

20. This summary is based primarily on interviews with Wade Hopping and Robert Rhodes, private attorneys, and Jake Varn, secretary of the Department of Environmental Regulation, in Tallahassee, Florida, March 26-27, 1980; see also Wade Hopping, "An Industrial Siting Act That Works: The Florida Electrical Power Plant Siting Act," a paper presented at the Eighth Annual American Bar Association Conference on the Environment, Airlie House, Warrenton, Virginia (May 4-5, 1979). Hopping's paper is summarized in *Environmental Law,* Newsletter of the American Bar Association Standing Committee on Environmental Law (Summer 1979), p. 1.

21. Hopping interview, March 26-27, 1980.

22. Florida Stat. 403.901 *et seq.* (1979). The act is discussed in Wade Hopping and Robert Rhodes, "Penetrating the Permitting Profligacy: The Industrial Siting Act of 1979," *Florida Bar Journal,* vol 53 (1979), p. 555.

23. For a discussion of the Washington act, see Auger and Zeller, *Siting Major Energy Facilities,* pp. 39-41; Temple, Barker, and Sloane, Inc., *Streamlining the Environmental Permitting Process: A Survey of State Reforms* (Washington, D.C.: U.S. EPA, Office of Policy and Resource Management, 1982), pp. 188-91.

24. See Auger and Zeller, *Siting Major Energy Facilities,* pp. 40-41.

25. For a fascinating account of this remarkable case, see Barrt M. Casper and Paul David Wellstone, *Powerline* (Boston, Mass.: University of Massachusetts Press, 1981); the case is also discussed in U.S. General Accounting Office, *Coal Creek: A Power Project with Continuing Controversies over Costs, Siting, and Potential Health Hazards* (Washington, D.C.: U.S. General Accounting Office, November 26, 1979).

26. Thomas Gladwin, "Patterns of Conflict over Industrial Facilities in the U.S., 1970-1978," *Natural Resources Journal,* vol. 20 (April 1980), p. 261.

27. Ibid.

28. Vesting a federal agency with power to preempt state and local environmental and land-use authority—in effect, forcing a state or locality to accept and bear the costs associated with an unwanted facility—would also raise constitutional issues that might serve only to delay crucial projects until answered. Although Congress generally has broad authority to regulate interstate commerce and, thereby, most major industrial projects, in several recent decisions interpreting the Tenth Amendment, which reserves all powers to the states not specifically granted to the federal government under the Constitution, the United States Supreme Court has held that "Congress may not exercise power in a fashion that impairs the states' integrity or their ability to function effectively in a federal system." *National League of Cities v. Usery,* 426 U.S. 833 (1976). Would a federal siting agency go too far if it overrode state and local objections to a project that

had serious fiscal and environmental consequences? Perhaps not, but with an increasingly conservative U.S. Supreme Court sympathetic to states' rights, opponents would be encouraged to litigate the issue. See also *Pennhurst State School and Hospital v. Halderman*, 101 S.Ct. 1531 (1981), holding that if Congress had desired states to comply with certain provisions of the Disabled Assistance and Bill of Rights Act it would have provided enough money for them to take steps to do so.

29. "Industrial Development Research Council Views on Environmental and Land Use Controls," *Industrial Development* (May/June 1975), pp. 7-8.

30. Ibid.

31. Auger and Zeller, *Siting Major Energy Facilities*, p. 27.

32. Ibid., p. 28.

33. Ibid.

34. Md. C.A. sec. 3-301 *et seq*. A case study of the act was undertaken as part of The Conservation Foundation's Industrial Siting Project. Angela Jordan, "Maryland Power Plant Siting Act of 1971—A Case Study," unpublished report (1979).

35. Association of Bay Area Governments, "Industrial Siting Pilot Project: Final Report of the ABAG Industrial Siting Task Force" (San Francisco, Calif.: Association of Bay Area Governments, February 1979); interviews with Charles Forester and Gordon Jacoby, Association of Bay Area Governments, Berkeley, California, February 23, 1979. Site banking and early site approval were also cornerstones of the Carter Administration's defunct proposal to speed the licensing of nuclear power plants. S. 2775, H.R. 11704, 95th Congress, 2d Session (1978).

36. For an excellent discussion of such controls, see Morris Sweet, *Industrial Location Policy for Economic Revitalization* (New York: Praeger, 1981).

37. Ibid., p. 43; discussions with David Hall, executive director, Town and Country Planning Association, England, various dates, 1974-82.

38. Sweet, *Industrial Location Policy*, pp. 45-46.

39. Telephone interview with Ray Brubaker, Dow Chemical Company, Sao Paulo, Brazil, April 23, 1980.

40. For a discussion of these and other regulatory reform proposals, see *Environmental Study Conference Weekly Bulletin* (March 9, 1981, and subsequent issues).

41. The Reagan initiative is discussed in greater detail in Lawrence Mosher, "Reaganites, with OMB List in Hand, Take Dead Aim at EPA's Regulations," *National Journal* (February 14, 1981), p. 256; and Shirley Scheibla, "Regulatory Relief," *Barron's* (June 8, 1981), p. 4. For an illustration of the executive order in practice, see Felicity Barringer, "Feud Tests OMB as Regulatory Watchdog," *Washington Post*, November 26, 1982. Aside from slowing the regulatory process, critics assert that the executive order allows industry to influence the substance of regulations through off-the-record contacts with the Administration.

42. The Bumpers amendment is discussed in Cornelius Kennedy, "The Bumpers Amendment: Regulating the Regulators," *American Bar Association Journal* (December 1981), p. 1639; and Francesca Lyman, "Tying up Regulation," *Environmental Action* (October 1981), p. 14. For a thoughtful legal analysis of the amendment, see Ronald Levin, *Judicial Review and the Bumpers Amendment*, a report to the Committee on Judicial Review of the Administrative Conference of the United States (Washington, D.C.: The Administrative Conference of the United States, 1979).

146 PART III: RESPONSES TO THE NEW REALITIES

43. Quoted in *Washington Post*, June 26, 1980.
44. Ibid.
45. This policy involves treating all stacks or sources of air pollution in a factory as though they were under a big dome or bubble, rather than treating each source individually. Industry would be allowed to add controls where they would be least costly as long as total pollution did not exceed a certain level. The proposed policy was recently struck down in court.
46. Statement by Barbara Thomas, commissioner of the Securities and Exchange Commission, quoted in *New York Times*, June 3, 1981.
47. Regulatory Procedures Act of 1981: Hearings on H.R. 746, before U.S. House, Committee on the Judiciary, Subcommittee on Administrative Law and Governmental Relations, 97th Cong., 1st sess. (March 24, 1981).
48. Milton Benjamin, "Plan Unveiled to Rejuvenate Atomic Power," *Washington Post*, October 16, 1982.
49. This description of the Colorado Joint Review Process is based on a series of interviews between 1979 and 1982 with Richard Lamm, governor of Colorado; Harris Sherman, former head of the Colorado Department of Natural Resources, now an attorney with the Denver office of Arnold and Porter; Monte Pasco, who succeeded Sherman as head of the Department of Natural Resources; Gary Fisher, former Department of Natural Resources employee and manager of the Colorado Joint Review Process; Dorothy Johnson, former Gunnison County, Colorado, administrator; David Leinsdorf, Gunnison County Council and resident of Crested Butte, Colorado; Michael Curran of the U.S. Forest Service; and several AMAX, Inc., personnel including Stanley Dempsey, Arthur Biddle, David Delcour, and Michael Rock.
50. For an excellent history of the Beaver Creek dispute, see William D. Johnson, "This Could Be the Last Resort," *Sports Illustrated* (December 15, 1979), p. 79.
51. Quoted in Ibid., p. 90.
52. The so-called Experiment in Ecology for the Henderson mine is detailed in Robert Cahn, *Footprints on the Planet* (New York: Universe Books, 1978), p. 71.
53. Address by David Delcour, vice-president of AMAX, Inc., National Conference on Coordinated Permitting Review for Energy and Mineral Resource Development, Keystone, Colorado (July 13-15, 1982).
54. Ibid.
55. Quoted in John D. Wiebmer, "A Clear Path through the Permitting Forest," *Mining Engineering* (July 1980), p. 791.
56. Stanley Dempsey and David Delcour, quoted in Lawrence Mosher, "A Fast Track for Colorado," *National Journal* (August 2, 1980), p. 1284.
57. Interviews with David Delcour and Arthur Biddle, Mount Emmons project manager, in Denver, August 1980, and later conversations with Biddle during 1981 and 1982.
58. Ibid.
59. Leinsdorf interview.
60. Dorothy Johnson interview.
61. Discussions with former Mayor W Mitchell in Crested Butte, Colorado, various dates, 1979 and 1980.
62. Curran interview.

63. Fisher interview.
64. Statement by William K. Reilly, president of The Conservation Foundation, at a press conference in Denver, April 10, 1980.
65. Address by Gary Fisher, Chevron Oil Shale Company, at National Conference on Coordinated Permitting Review for Energy and Mineral Resource Development, Keystone, Colorado (July 13-15, 1982).
66. Address by Thomas Gladwin, associate professor, Graduate School of Business Administration, New York University, at National Conference on Coordinated Permitting Review for Energy and Mineral Resource Development, Keystone, Colorado (July 13-15, 1982).

Chapter 6

Responses to the Real Problems: Permit System Innovations for the 1980s

There are four outstanding problems associated with industrial siting:
- The regulatory system's confusing structure needlessly lengthens the permitting process, particularly when multiple permits from multiple agencies are required for a project.
- The system creates uncertainty, which plagues industry. Project planners often cannot accurately predict whether a project will be approved or when.
- The system does not always produce environmentally sound decisions.
- Decisions often lack finality, since administrative decisions can be reversed or challenged in court. This further lengthens the process and adds to the uncertainty.

Several culprits, of varying importance from case to case, can be identified:
- Lack of intergovernmental or interagency cooperation and coordination.
- Failure by project proponents to evaluate environmental impacts and involve the public from the start.
- Absence of information that both regulators and project proponents need to make sound decisions.
- Inadequate understanding by regulators of industry project planning needs and by project planners of the constraints facing regulators.
- Lack of a predictable timetable in the review process. The rules of the game can change at any time, even after a permit is issued, and regulators sometimes have no impetus to say yes or no quickly.
- Increasing unpredictability and open-endedness in the judicial review process.

"Quiet" reforms that focus on procedural and institutional changes offer the most promise of effective, long-lasting improvement in the siting process. Indeed, the current search by some for cure-alls such as those discussed in chapter 5 may actually set things back, as promising, less-visible efforts are swept aside in the rush to reform.

This means more than just tinkering, however. Some fundamental changes are necessary in the way regulators handle project reviews and in the way corporations plan major facilities.

THE ROLE OF GOVERNMENT

Inevitably, when people talk about reforming the siting process, they focus on governments as prime candidates for change. Overlapping agency requirements and lengthy permit reviews are seen as problems that originate within government. Indeed, to solve many deficiencies in the system, government must reform.

Coordination through Better Management

Federal, state, and local agencies are already experimenting to see if the siting process can be improved by using everyday business management techniques.[1] Most of these experiments are being conducted without specific statutory authority and do not create new bureaucracies, add new regulations, or preempt existing laws.

Measures to improve coordination must be well thought out and implemented if they are not to be counterproductive. The most elementary approach to "better coordination," for example, has simply meant referring a proposal to more outside agencies and bodies, and then sitting back and hoping for the best. Although such informal referrals can be valuable, they can also slow the permitting process without achieving better reviews. If an agency to which an application is referred for comment has no commitment to the process, it often will "sign off" with little or no review, particularly if it already has a heavy work load and is not accountable for a final approval or rejection.

Today, several new approaches are being used to help coordinate the management of industrial siting, thus improving its effectiveness and efficiency. Among the most promising management techniques are designation of lead agencies, development of expert project teams, convening of joint or consolidated hearings, use of agency project managers and regulatory escort services, and reliance on interagency mediating bodies during the regulatory review process.

Lead Agencies

The lead/cooperating agency approach is a simple one that has already paid dividends for several major projects. The federal Council on Environmental Quality (CEQ) made it a key element in its ambitious attempt to overhaul the National Environmental Policy Act (NEPA) environmental impact statement (EIS) process in 1978.[2] Interestingly, the prime mover behind the reforms at CEQ was a participant in the Dow siting controversy who drew on that experience in revamping EIS procedures.[3]

In the past, squabbles over which agency should take the lead in writing the EIS for a project often led to delays. The same was true at the state level. To tackle such problems, CEQ proposed that the federal agencies involved in a project decide which should act as the lead by considering how five factors (listed in descending importance) relate to each agency: magnitude of agency involvement; project approval/disapproval authority; expertise concerning the project's environmental effects; duration of agency involvement; and sequence of agency involvement.[4] Should a deadlock persist, the agencies must move within 45 days to have CEQ appoint a lead. CEQ then has 20 days to respond.

Once selected, the lead agency supervises EIS preparation, identifies all other permits and reviews needed before construction on a project can begin, and brings other agencies into the review process. Moreover, instead of merely allowing the lead agency to refer a completed EIS to other agencies, the CEQ regulations emphasize early interagency cooperation by designating other interested agencies as "cooperating agencies" with a formal role in the process.[5] According to one CEQ official: "Now an agency other than the lead agency with jurisdiction or expertise will not just sit back and criticize somebody else's EIS. As a 'cooperating agency,' it will help write it in the first place. This too will produce a better draft document and will reduce the need for delays later on."[6]

Several states have taken the lead agency concept a step further by integrating it with the permitting process. Thus, agencies can save time by using information produced for the EIS rather than developing additional information separately. The CEQ regulations also encourage other permit reviews to run concurrently with EIS preparation.

Another good example of how a vigorous lead agency can improve the process comes from the Bureau of Land Management (BLM) in the Department of the Interior. BLM's Office of Special Projects (OSP)[7] played an important coordinating role in the permitting of several major energy and mineral developments in the late 1970s. Many energy and mineral projects in the West involve BLM land, either directly or through pipeline crossings. OSP worked closely with BLM state directors to divide authority for projects and to establish a single contact in key Interior offices who agreed to take responsibility for keeping things moving. OSP also involved other federal and state agencies, signing memorandums of agreement delineating project-review responsibilities, and then monitoring progress to make sure there was no slippage.

The lead agency idea is not without pitfalls and must be implemented with care. For example, according to BLM officials, OSP ran into some problems in its lead-agency function because state directors within the bureau complained that OSP was invading their home turfs and acting too independently. A 1981 internal study of BLM permitting procedures recommended that OSP's role be defined by a BLM policy statement that would be widely circulated within the agency.[8] Soon thereafter, the Reagan Administration dismantled the office.

Expert Project Teams

An extension of the lead agency approach is organizing a project team within an agency or among agencies to take responsibility for a major facility. Again, the theory is that, instead of relying on interagency referrals, project review proceeds more smoothly and more thoroughly if regulators meet face-to-face with each other and maintain close contact with the applicant.

This practice has increased in the last few years and has been implemented in some of the programs discussed in chapter 5. Georgia uses a project-team approach within its Environmental Protection Division; and the state of Colorado and the U.S. Forest Service, under the Colorado Joint Review Process, organized a project team to handle an application to build a large molybdenum mine.

Most permitting bodies, even ones with adequate personnel and resources, gear up slowly to handle major developments. Months can be lost if agencies wait until an application is submitted before organizing to deal with it. Instead, agencies should organize environmental "SWAT" teams that can take action as soon as a project proponent lets the agency know what is planned—a notice that should come as early as possible.

BLM, drawing on what it learned in dealing with several major energy projects, developed the OSP environmental impact team so that it could stay on top of any developments. Once BLM received notice of a major project, it could quickly call together previously designated, experienced team members. In its first few trials, the special team dramatically reduced confusion and delay in project evaluation and produced better environmental impact reviews.[9]

The project team approach was used among agencies to review a proposed U.S. Steel facility on rural land in Pennsylvania and Ohio along Lake Erie.[10] Instead of simply appointing a lead agency to develop an EIS for circulation among all interested parties, as was done in the Dow case, an executive team and a separate technical team were organized to oversee preparation of a comprehensive environmental impact analysis to meet federal and state requirements.

The members of the executive team were high-level officials drawn from federal agencies—including the Army Corps of Engineers, the regional Environmental Protection Agency (EPA) office, and the Fish and Wildlife Service—and from the Ohio and Pennsylvania governors' offices. A vice-president of engineering sat in for U.S. Steel. The technical team consisted of representatives from the same organizations plus U.S. Steel's EIS consultant. Representatives from other federal agencies, such as the Department of Commerce, were added later in the process.

The technical team's first job was to identify significant environmental issues and critical decision points in the evaluation process. Working groups were set up to establish baseline data in relevant areas (air, biota, solid waste, socioeconomics, and water); as the team proceeded, traffic and archaeological sites were also identified as major issues and evaluated.

In this first phase, the technical team held a series of workshops to get public comment and gather baseline information. Next, U.S. Steel presented a conceptual design for the facility. The technical team then evaluated environmental impacts based on the proposal and baseline data, using those impacts as a starting point for establishing mitigation measures.

Throughout the process, the executive team's major role was to maintain cooperation among members of the technical team. The feeling was that the continuing attention of high-level officials on both sides would keep reviews moving and would help feed environmental factors into the project design process before plans were submitted. Time was also saved by undertaking *parallel* rather than *sequential* reviews and by assigning responsibilities according to a mutually agreed on timetable. As a result, the draft EIS was produced in 16 months—"akin to a two-minute mile," in the words of one participant.

The process did have its shortcomings, however, and the final EIS approval was later challenged in court.[11] Local governments had no official representation on the technical committee. Likewise, although states were represented through their governors' offices, state agencies with greater technical expertise were not directly involved. Some participants in the EIS process suggested that it should have been designed to coordinate better with project permitting. These two processes were kept apart on the theory the EIS involved legal and public relations considerations separate from the highly technical issues involved in permitting.

U.S. Steel's EIS consultant also recommended changes to improve the process. He suggested establishing a longer, more realistic schedule initially (18 months) and designating the EPA rather than the Corps of Engineers as the lead agency because of its broader point of view and expertise in various environmental areas.

Economic conditions caused the U.S. Steel project to be shelved. The project EIS did receive government approval very quickly, however, demonstrating the value of a cooperative project-team approach. Refinements of the approach—for example, closer coordination of permitting with EIS preparation and inclusion of local governments—could perhaps help to avoid legal challenge.

Joint or Consolidated Hearings

A useful, fairly simple adjunct to the lead agency idea is the use of joint or consolidated hearings. Many states—California, Florida, Oregon, New Hampshire, and Maryland, to name a few—already have either statutory or informal procedures allowing joint hearings, which often have been very successful.

Under the Florida Power Plant Siting Act (see chapter 5), for example, supporters say not only that giving each proposal a single comprehensive hearing has helped cut permitting time, but also that the procedure has improved the quality of decisions. An attorney who has represented utilities in the process says: "It cuts through the baloney. It also gets better results. A decision about air emissions may somehow affect water-pollution control. The staffs sometimes don't realize this. The hearing process tends to pull things together."[12]

The Florida approach and other successful joint hearing initiatives have several common attributes. Each hearing is preceded by extensive informal contact and negotiation among the project proponent, government, and the public. An independent hearing officer or an agency that is perceived to be neutral and fair oversees the process, with responsibility for all notices and coordination of appearances. Finally, the scope of a hearing is defined closely beforehand, and the hearing is conducted accordingly.

If joint hearings are not handled sensitively, however, they can make things worse. Joint hearings must be held early in the process and be mandatory—or, if voluntary, offer clear benefits—if they are to have a good chance of being used. In the Dow case, for example, the joint hearing came too late in the process, after minds had been made up and battle lines drawn. Dow had, unfortunately, resisted an offer by California officials to hold hearings eight months earlier, preferring instead to deal with each agency separately.

The state of Washington's Environmental Coordination Procedures Act, discussed in chapter 5, provides another example of how joint hearings may not improve the siting process. Many project applicants have used the first step in that law to identify necessary permits, but they have then dropped out of the process, in part to avoid going to a joint hearing. They fear that the hearing will be too broad in scope, allowing objectors to raise issues unrelated to necessary permits. Some industries also worry that getting two or more agencies in a room together may raise some questions that might not otherwise arise.

This problem of a perceived scattered focus can be dealt with through guidelines that clearly define each hearing's purpose. If industries want the process to be more efficient, they must be willing to put their cards on the table and to support better project reviews. Progressive, well-prepared firms generally should have little fear of joint hearings. As one attorney, who represented a uranium mining firm at a joint hearing in Colorado, observed: "We liked it because it cut down on the meetings involved and made the environmental interests present their arguments at one time. It also gave the agencies involved a far better understanding of the project, which helped us in the long run."[13].

Some environmentalists also view single joint hearings with skepticism. The meetings are often lengthy and very formal, necessitating the hiring of attorneys and costly preparation. Moreover, a single hearing means, in the words of one environmentalist, "that we get only one bite at the apple.".

Multiple hearings give project opponents time to build a case and provide more opportunities for legal challenge. However, just as industry must accept the risks of joint hearings in exchange for increased efficiency, so must environmentalists balance their desires for additional time with better project evaluation and less palatable alternatives that may be waiting in the wings.

Project Managers and Escort Services

Agency permit contacts and regulatory escort services that introduce applicants to regulators and that help identify feasible sites can also help the system run more smoothly. Several states are already giving each major proposal a single contact within each agency; the contact is responsible for keeping the applicant informed about the progress of its permits.

Permit contacts are, in effect, the project managers for their agencies. To be most effective, they should be assigned to a project for its duration and have authority to commit their agency to certain actions quickly without drawn-out review by superiors. For example, these managers might agree to have their agencies share the costs of analyzing an environmental problem that surfaces late in project review.[14]

Some states have gone beyond providing permit contacts to establishing what might be called industrial escort services. These services put project planners in touch with agency personnel who have a good working knowledge of what environmental and land-use issues a site might raise.

Industries coming to Georgia, for example, usually make their first contact with the state's highly regarded Department of Industry and Trade (DIT), which helps identify sites that meet a firm's particular locational criteria.[15] If the DIT staff senses that a new plant may have pollution problems, the Georgia Environmental Protection Division (EPD) is contacted for information on air, water, and land-use resources. (The EPD was discussed in detail in chapter 5.) Because the EPD has authority over a wide range of environmental programs, from air quality to solid-waste management, it has a good sense of whether a particular site will have problems. Officials say this early contact helps steer industry from areas of the state with environmental problems: "We didn't try to fit a square peg in a round hole; EPD pretty well dictates where a plant can go from a pollution point of view." This screening process has won plaudits from environmentalists as well as industry.

In the wake of the Dow debacle, California organized an Office of Permit Assistance (OPA) that provides similar services.[16] OPA organizes preliminary meetings with key agencies and project proponents, coordinates preparation of joint environmental documents when both federal and state agencies are involved, and convenes joint hearings when several agencies are concerned.

Interagency Mediating Bodies

Another related, but more sophisticated, technique to coordinate management is the creation of a dispute-mediation agency so that, for example, an applicant does not get confusing signals about what a government expects from it. In the SOHIO dispute, one state agency preferred a coastal site to avoid air-pollution problems while another urged SOHIO to locate inland to avoid adverse land-use impacts in an undeveloped area. A mediating body might have resolved this mixed signal.

The federal CEQ offers an example of how an agency mediator can help. Assume a company applies to the Corps of Engineers for a dredge-and-fill permit as part of its plan to build a petrochemical facility on the coast. The Corps, which has broad discretion to issue or deny the permit in the public interest, circulates the application to other federal agencies, such as the EPA and the Fish and Wildlife Service, as required by law. If all agencies agree there will be no adverse impact, usually the project will receive quick approval. But what if the Corps says "go" and the Fish and Wildlife Service says "no," worrying that dredging and filling could destroy duck nesting areas or spawning grounds?

Statutorily, if a dispute arises at the first level of review, the decision is kicked to a higher level in the Corps, but this process can be time-consuming, particularly if positions have hardened. This is where CEQ fits in, since it can act as a neutral convenor to get the warring sides to talk with each other. CEQ's evenhanded mediating often has been instrumental in breaking up regulatory log jams blocking projects (for example, by refereeing a dispute over impact data or helping explore mitigation alternatives). A comprehensive study of how federal agencies respond to NEPA found that: "One of the most significant recurring themes was the perception of CEQ and the NEPA process as mediating institutions. Most of this activity is informal and never becomes public."[17]

California's OPA functions very similarly to CEQ. If a dispute between state agencies—say, the coastal commission and the state air board—arises during project review, OPA can convene the parties and attempt to work out a position that all can live with. OPA has other powers that help avoid disputes. It encourages federal and state agencies to prepare joint environmental reviews of a project; resolves interagency disputes over lead agency status; coordinates preparation of interagency memorandums of agreement, which delineate who will do what and when in the review process; and convenes consolidated hearings on behalf of other agencies. Projects in which it has been active run the gamut from large energy proposals—notably refineries and offshore drilling—to mining and geothermal exploration.

The Need for Predictability

Even without regulatory hitches, most major industrial, energy, and mineral projects take years to move from the planning phase to full-scale operation. The timing of a project is crucial, often linked to meeting a particular market at a specific time. Any major deviation in the project schedule spells trouble. Sometimes rules and regulations change midproject, forcing a proponent to redesign a facility. In other instances, regulators are slow in issuing permits, not for substantive reasons but simply because they are under little pressure to act expeditiously.

Contrary to the common perception, regulators possess great flexibility in applying many environmental-protection laws. In the Hampton Roads case discussed in chapter 2, for example, the governing standard for a federal dredge-and-fill permit was whether, on balance, granting the permit would be in the "public interest," a nebulous concept that lacked legislative definition. Unfortunately, there is little likelihood that legislatures will enact laws that eliminate—or even reduce—agency discretion. The regulatory issues are too complex, and it is often impracticable for a large group of legislators to agree on specific and often technical standards. Moreover, even where fairly precise standards are established by law or administrative policy, leeway exists in applying them in a specific case.

The real source of predictability will be the experience that both regulators and industry gain as they carry out the industrial siting process. Already, there are signs of improvement from across the country. The preparation of NEPA-required EISs, for example, has improved considerably, particularly since CEQ's 1979 reforms; SOHIO used its Pactex experience to site a major chemical facility in Texas successfully.

Project Decision Schedules

A permit application may become bogged down almost as soon as it is filed. The regulatory agency requests more information before it will process the permit. And then some more. And more. Time is wasted because the company leaves out information that the regulatory agency thinks it needs to evaluate the application; yet the agency often does not give the applicant a complete list of information needed. The problem may be exacerbated if the agencies have no firm deadlines for reviewing specific permits once an application is complete.

Many in industry and in legislatures believe that one answer to problems of permit delay is mandatory decision schedules. Under one version of the energy mobilization board (EMB) proposal, for example, a 12-month limit was set on decisions to be made by federal, state, and local agencies. If the deadline was not met, then the EMB could either go to court to enforce the deadline or step in and make the decision itself. Some states have already adopted this type of approach. In California, a law enacted in response to the Dow case provides that if state agencies do not act on an environmental-permit application within 18 months, the project is deemed to be approved.[18]

Although mandatory or expedited decision schedules may instill some discipline in the permitting process, they create their own problems. Few people like the idea of a project being approved by default. Instead, a government agency facing a deadline may simply reject an application unless its proponent agrees to an extension. A 1978 report on ways to improve Maine's environmental-permit system concluded that rigid timetables in other states had not increased certainty in the review process, since deadlines were frequently extended or simply ignored.[19]

Moreover, an across-the-board time limit for all projects hardly reflects the great differences between an oil pipeline and a big petrochemical facility. Six months may be more than enough time for a pipeline but far too little to evaluate a major petrochemical refinery. Governors expressed concern about this problem in commentary on proposed EMB legislation:

> They . . . fear that the [EMB] will be insensitive to unavoidable delays such as the failure of applicants to submit required information in a timely manner, availability of new information affecting the siting decision, and the time and resources necessary to adequately review the complex technical details of a major energy proposal.[20]

Finally, poor project evaluation caused by hurried reviews would undoubtedly increase litigation by opponents.

There are more promising ways to make the process more efficient. Several states now require that agencies make a "completeness determination" within a prescribed time after the submission of an application.[21] Once the deadline passes, the agency is foreclosed from asking for more information or requiring additional monitoring. Care must be taken in applying such deadlines, and provisions must be made for cases where an unexpected environmental issue pops up later in the review process or where the applicant is dragging its feet in supplying information. But overall these deadlines can help start the process in a timely fashion.

Once a completeness determination is made, voluntary decision schedules have proved far more valuable than mandatory ones (such as EMBs') in keeping project reviews on time. Inflexible permit deadlines usually apply to only a few permits or one level of government and are not set for the overall project approval process. Thus, simply forcing issuance of a few permits does not guarantee a speedup.

Voluntary schedules can be tailored to specific projects. Increasingly, this is being done with a formal agreement signed by a project proponent and regulatory agencies, whereby all parties agree to use their best efforts to make decisions according to a predetermined schedule. In this way, permitting is viewed realistically as an overall process rather than from a narrow permit-by-permit perspective. In addition, this flexibility appears to breed a real sense of cooperation among government regulators and project proponents, while binding schedules seem to promote an atmosphere of confrontation.

The Colorado Joint Review Process (CJRP), detailed in chapter 5, is an excellent example of this tailoring. The state has published, in consultation with industry, model decision schedules for major energy and mineral projects, but they are illustrative only. When an industry volunteers to participate in the CJRP, a specific schedule is drawn up for that project.

A formal schedule was also used in the permitting of the proposed Alpetco oil refinery in Alaska. In 1977, Alaska and Alpetco signed an agreement in which the company agreed to have all permit applications and environmental impact analyses completed by a specified date. In return, the state agreed to rule on the application within a certain time. Although that project was shelved because of a drop in demand for refined oil products, reviews for the project stayed on schedule.

At the federal level, CEQ's revamped NEPA regulations also promote project decision schedules. BLM's OSP used decision schedules in writing impact statements for a variety of projects on federal lands. OSP not only established a schedule for review within the Department of the Interior but also coordinated review by other agencies, such as the U.S. Forest Service.

Project Tracking Systems

A project tracking system keeps tabs on an application as it proceeds through the permit process. Such a system is especially useful in conjunction with voluntary decision schedules. The rationale, again, is that if permitting agencies know that someone is watching and will hold them accountable if reviews are needlessly bogged down, then they will process permits more expeditiously.

Voluntary decision schedules work best when monitored by a single office or agency. In Colorado, a special office at the Department of Natural Resources (DNR) acts as a project monitor. Within BLM, the OSP, which had direct lines into the director's office, kept track. DNR does not have nor did OSP have formal power to enforce decision schedules, but a call to an agency head does wonders to unstick a bogged-down permit review.

In 1980, the U.S. EPA adopted the so-called PEP (priority energy project) system, a very ambitious project-tracking effort that grew out of the ill-fated EMB legislation.[22] Prompted by complaints that its permitting process was delaying needed developments, the agency hoped to monitor the permitting and review progress of major energy facilities across the nation, focusing on EPA permits.

The PEP system was very simple in concept and practice. Headquarters staff within the Energy Facilities Branch checked on the

162 PART III: RESPONSES TO THE NEW REALITIES

status of federal reviews of almost 100 major energy facilities every six weeks and sent a report to the EPA administrator and other agency personnel staff. Included were EIS reviews and permits under the Clean Air and Clean Water acts, the Resource Conservation and Recovery Act (RCRA), and the Toxic Substances Control Act. If projects seemed to be stalled, the system could identify problems and resolve them quickly. In addition, PEP promoted the setting and meeting of deadlines for reviews. Finally, EPA headquarters could make expertise and advice available in cases where reviews were being delayed on technical grounds.

By the end of 1982, the PEP system had compiled an impressive record: on projects for which EPA had set decision schedules, regional offices had met or beaten over 80 percent of the target dates, with an average processing time of less than nine and a half months.[23] Because of the system's success, EPA has now established an agency-wide permit-tracking system to be administered in each region. Each region will have a coordinator who will act as a contact point for industry. Seven measures will be used to evaluate permitting performance: average processing time; response time and success in getting complete applications; percentage and extent of permit actions exceeding target maximum; percentage of permit actions exceeding negotiated decision schedules; number and proportion of pending complete applications exceeding target maximum processing time; progress against targets for processing existing source permits; and milestones for key program action. If these measures are not satisfied, the regional EPA administrator will be held accountable, according to headquarters officials.[24]

Grandfather Clauses

Uncertainty in the permitting process is created when laws or regulations change after a project is under way. Some of these changes may be entirely justifiable on public health grounds (for example, enactment of hazardous-waste-disposal controls). Hardly anyone would argue against applying a new law critical to the protection of public health even if a facility is under construction. But that is the easy case. The impetus behind other changes in land-use controls may be less compelling. For example, local land-use laws are sometimes enacted in response to protests by citizen groups and nearby homeowners concerned about their property values.

The problem of mid-project legislative revisions should be substan-

RESPONSES TO THE REAL PROBLEMS 163

tially reduced in the 1980s, simply because many of the necessary laws are now in place. Still, changes in standards and regulations under existing laws remain a distinct possibility; new standards for specific problems like acid rain are also in the offing. Furthermore, local authorities will undoubtedly exert control over proposed facilities by changing zoning regulations. State law usually allows such changes until construction actually begins, even if industry has spent millions on planning.

Projects can also be delayed when regulators decide to relax laws. For example, some of the confusion over implementation of the hazardous-waste-disposal provisions of RCRA is due to the Reagan Administration's decision to rewrite what it says are overly strict rules.

Grandfather clauses allowing industries to proceed despite changes in requirements will almost certainly play an increasingly important role in the future, as regulators attempt to solve this problem. The question is where to draw the line. One early version of the EMB legislation would have protected any energy project from future regulation once substantial commitments had been made for design, site clearing, and machinery.[25] The result would have been to exempt several synfuels projects from most existing clean air and water requirements. Such an extreme approach has little to recommend it and would have engendered heated opposition.

Various other approaches to grandfather clauses need to be tried. The Clean Water Act exemplifies one option—a clause protecting a new plant from any tightening of source performance standards for ten years, unless public-health reasons mandate otherwise.[26] The provision applies to facilities already built or under construction.

But what of cases involving multiple permits, when construction cannot begin until all approvals are in hand? Say an industry gets several permits, but decides against buying land or starting construction until all the permits have been received. If the regulations underlying an already issued permit change, should the firm receive grandfather treatment for the millions of dollars it has put into design and engineering work, as well as into land purchase and clearance, or for legal obligations into which it has entered (such as contracts for production equipment)? One commentator has suggested that the answer is to allow vesting in such cases:

> Thus, for example, the award of a permit to construct by an AQMD [air quality management district] could possibly be vested by execution of a binding contract for purchase of the boilers or other equipment, and construc-

tion of foundations to install such equipment at the site. The vesting issue should be considered independently as to each permit and, based on the likelihood that retroactive requirements will be imposed, the company can make a case-by-case decision as to whether the capital investment necessary to acquire a vested right as to a particular permit should be made.[27]

This approach seems cumbersome as applied to big projects requiring multiple permits, and it seems extreme in granting protection even where health issues are involved. A more palatable approach—suggested by a recent study of the so-called vested rights issue in a land-use context—might be to give projects a vested right to proceed despite changing land-use laws if the project proponent has made substantial planning and design expenditures (but has not started construction) based on already existing laws, unless there were compelling public-welfare considerations to the contrary.[28]

The 1979 Florida Industrial Siting Act suggests another interesting variation.[29] Once a local government has given zoning or other approval, it cannot change ordinances, plans, or development orders that affect the project for two years unless the applicant concurs or a comprehensive amendment is made to its local land-use plan. This provision aims to stop local governments from making changes in local zoning or land-use laws directed specifically at an already-approved major facility. The act also gives industries some protection against changes in state laws. A project must comply with new environmental laws or regulations only if those rules apply to existing projects similar to the one being certified. In effect, a certified facility is considered to exist even though construction has not been completed.

Another possible approach is to insulate projects that have received permits from any change unless the regulatory authority can prove by a preponderance of evidence that significant health impacts necessitate retroactivity. This would, in effect, switch the burden of proof, normally placed on the party challenging a regulatory action, to the regulators.

Perhaps the most interesting experiment with a grandfather clause, one that may serve as a prototype for other environmental laws, involves the federal Endangered Species Act.[30] Recent amendments to the act provide that, if a project proponent who finds an endangered species on a proposed site adopts a plan approved by the U.S. Fish and Wildlife Service to protect and ensure the existence of that species, the project cannot be stopped just because it may have some incidental impact on the species. Moreover, once that plan is accepted, the project proponent is given long-term assurances that the federal government will not impose new mitigation requirements.

As one observer has commented on this approach, "The new amendments to the Endangered Species Act are oriented towards attempts to work out solutions to environmental problems in a constructive manner by providing the development industry with long-range assurances in exchange for cooperation."[31] Although it is too early to tell whether this will work well enough to be considered as a model for changes in other laws, it certainly bears watching.

Judicious use of grandfather clauses in combination with other techniques should reduce uncertainty in the present system without ignoring laws crucial to protecting public health and important environmental values. However, grandfather clauses are not substitutes for other processes, such as lead agencies and mediating bodies, that can improve the quality of reviews as well as speed them up. Moreover, grandfather clauses cannot work without coordination among agencies. For example, even if a project is safeguarded under federal law by a grandfather clause, the proposal may still be subject to changes in *local* law. Close and continuing contact among agencies remain essential.

Regulatory Personnel Problems

A frequent complaint from industry is that regulators do not always understand the needs of project planners. As one businessman has put it, "Very few civil servants are familiar enough with the industries they regulate."[32]

Some regulators, particularly in smaller agencies or governments, do lack the training or experience necessary to deal with major industrial projects. A regulator who does not fully comprehend a project may be hesitant to approve it or may make demands that industry has difficulty meeting. Frequent requests by regulators for detailed design and engineering drawings prior to approval are a prime example. For industry to produce such detailed information for big projects would be costly and risky when final approval is not guaranteed.

The problem of regulator expertise is particularly acute when a new process or technology is involved, but similar problems can arise even with "standard" production processes when firms try to site projects in regions with little industrial development. While Texas environmental regulators may know a great deal about petrochemical refineries and energy projects, for example, agency personnel from Alaska or Utah may feel less assured. On the other hand, the regulator may feel so overawed, overwhelmed, or overworked that approvals are given without adequate environmental analysis.

These personnel problems have several sources. The regulatory system is new, and regulators, just like corporate project planners, are still learning the ropes. After all, only a handful of petrochemical facilities may be built in this country over a five-year period. It is not surprising that regulators must learn from experience. The SOHIO and Dow cases are painful examples.

This situation has been exacerbated by a shortage of review personnel in many agencies and by a turnover of personnel as people either move up in government or on to higher paying jobs in industry. To cite an extreme example, during 1980 the Puerto Rico Environmental Quality Board lost 37 of its 50 engineers.[33] In Colorado, the two top officials responsible for managing the state's expedited permit process left for jobs in private industry less than two years after their state positions were created.

Within environmental or land-use control agencies, project permitting is often considered to be less prestigious work than are long-range planning, standard setting, and enforcement. As a result, "rookies" are often assigned to permit reviews, forcing industry to deal with people who are not especially knowledgeable about how a particular process works or about the dictates of project planning. Once a rookie becomes seasoned, that person often moves to another position, which starts the learning process over again. Until such positions are staffed with experienced, adequately paid people who can take pride in their work, there is little hope for significant improvement.

These problems may become more significant as responsibility for environmental programs shifts from the federal government to state agencies. Recent statistics show that most states are slowly cutting back on regulatory personnel and that salaries are falling farther and farther behind the private sector, particularly as federal aid drops.[34] Moreover, local governments facing tax limits similar to California's Proposition 13 are also reducing the number of land-use regulatory personnel.[35]

Even the federal EPA has cut back drastically since 1980, despite increasing responsibilities in areas like hazardous-waste disposal. In 1982 the Reagan Administration tried to halve the number of EPA personnel. Although the proposal was not completely implemented, in 1983 the Administration suggested even deeper cuts, but relented under congressional pressure.[36] These trends are ominous both for industry concerned about speedy reviews and for conservationists concerned with adequate environmental impact analysis and protection.

In the past, the federal government has provided not only financial assistance to state regulatory agencies, but also technical advice for the staff. However, the immediate prospects for increased federal aid are bleak, and it is uncertain whether state legislatures will vote funds to pick up the slack. Early indications are not encouraging.[37] For example, when the Department of the Interior announced it was zero-budgeting funds for state historic preservation offices that carry out federal preservation mandates, including important resource surveys that can help developers avoid problems, several states quickly announced that their offices would close. Congress refused to cut the funds completely, but aid has been reduced significantly.

Perhaps some in industry view these cutbacks as a way to reduce regulation across the board, but the underlying laws, often with citizen-enforcement provisions, remain on the books. As a result, regulators are likely to proceed even more slowly as their workloads increase. Legal challenges to permits, claiming inadequate review, may also increase. Either way, industry loses.

One industry official has observed: "It does little good to lash out at government inefficiency if it does not have what it needs to work with. The thing to do is go and help the government get what it needs."[38] Another industry representative has said that the right approach to "rookie syndrome" is for industry to do *all* the homework for the regulator so that decisions are technically and legally sound. Some smart firms are following these suggestions. In addition, some companies are providing funds for local governments so that they can adequately plan for the impacts of major industrial projects. There are also a few examples at the federal level of industry lobbying to ensure that certain environmental programs—such as hazardous-waste-disposal regulations—are implemented quickly to remove uncertainty about how and when new laws will be applied.

If industry does not provide this sort of assistance, state and local governments may turn to permitting fees to support regulators in an era of tight money and defederalization.[39] Such a move has ample precedent. For example, many local governments already charge major fees for housing development proposals to defray permitting costs.

Indeed, there already is some movement toward charging substantial fees to process and review applications for environmental approvals of industrial projects. In Florida, if a company should elect to follow permitting procedures of the state's Industrial Siting Act of 1979, it would be required to pay a sliding scale of fees ranging from $2,500 to $25,000 to defray the cost of proceedings.[40] New York's Uniform Procedures Act, which deals with administration of state environmental regulations, authorizes charging applicants a permit fee of up to $500, plus the costs of any public hearings.[41]

If it is assumed for the moment—and it is a major assumption—that the personnel squeeze can be eased, how can regulators become knowledgeable about the industries they control? Colorado's series of model decision schedules for specific types of plants illustrates one way to improve the present situation. The schedules show regulators how corporate project planning will proceed and identify benchmarks in the regulatory process.

Industry itself offers another approach worth considering. Various organizations (often industry groups) have sponsored workshops and seminars, led by experienced industry people, to discuss such topics as the steps in building a synfuels plant. Unfortunately, budgetary constraints mean that a dwindling number of government officials can attend. The programs are costly—often $500 to $1,000 per person, in addition to travel and lodging expenses.

Industries, especially trade associations, could help solve the regulators' knowledge gap by helping finance their attendance at such meetings, perhaps by providing them with scholarships to attend such meetings or by holding a yearly informational gathering for appropriate regulators. To increase the sessions' effectiveness, they might be cosponsored with or run by a "neutral" organization and kept as free as possible of public relations material. Despite undeniable risks, the proceedings of such meetings, explaining the ins and outs of building a particular type of plant, would be extremely valuable to regulators when they returned to their jobs.

THE CHALLENGE TO INDUSTRY

Many recent initiatives to improve the siting system have looked to government for all the answers and have overlooked industry's part in the regulatory process. But regulator and regulated must work together; industry, as well as government, can act to expedite permitting. One aspect of that task, helping regulatory agencies compensate for personnel problems, has already been discussed, but there is much more that industry can do. By learning from past mistakes, companies not only can save time and increase the likelihood their projects will be approved but also can proffer better environmental impact analyses and mitigation measures.

The secretive, behind-the-scenes method of siting facilities that was commonplace in the 1960s clearly will not pass muster in the 1980s. William Ruckelshaus, EPA administrator, sees such change as essential to corporate success. "The issue, quite properly, is not avoiding environmental regulations, but adapting to them. Ultimately, the fact that one corporation is more successful than another will depend on which one understands the origin of changes mandated by society at large, and successfully adapts to those changes."[42]

Industry will not make that adaptation overnight or without exasperation. Some in industry view changes in corporate planning procedures warily because they seem to promise more delay. Those businesspeople should bear in mind that speed at the start of the permitting process does not necessarily indicate a project's success. In the Dow case, for example, unseemly haste in the early stages helped to fan opposition and, later, to kill the project. By contrast, a company that devotes much time during the early phases of a project to negotiating with regulators and allaying public apprehension may cut the total time it takes to acquire permits. Once one or two key permits are approved, the remainder of them follow fairly quickly.

Despite any recalcitrance some in industry may have, their businesses must adapt to environmental regulation if they are to survive. They can meet this challenge by taking the initiative and showing the public they recognize, and want to solve, environmental problems. Early and complete environmental assessments, improved project organization, a renewed commitment to public involvement, and adoption of innovative dispute settlement could help them in the efforts.

Early Environmental Impact Assessment

A key element in predicting whether a project will be blocked or seriously delayed is whether the project proponent—either a company or an agency—evaluated environmental impacts early in the process. Unfortunately, early analysis usually is weak or nonexistent. Delays result as regulators demand additional information or companies change their plans in response to objections by regulators and project opponents.

A 1974 environmental planning survey of major projects by multinational corporations found that most were "considering little beyond immediate economic and technical aspects" in project planning.[43] Since then, some improvement has been made. A few companies are, in fact, quite good at evaluating a project's environmental consequences, particularly air and water pollution, *before* they settle on a site. Even more are becoming fairly sophisticated in the way they evaluate impacts *after* a project has been approved internally and a site has been selected, and many are much better at satisfying objections to a project by taking action to mitigate adverse impacts. One high-ranking executive noted: "Ten years ago I did not know what an estuary was. We have since learned a lot about estuaries and wetlands and found out that if you destroy mangroves, it costs a lot for maintenance dredging in the harbor. It really makes good sense to be involved in environmental assessment and good planning."[44]

On the whole, however, environmental and land-use considerations still do not figure into early project planning and site selection as much as do factors such as transportation and labor costs. "Environmental executives" tend to be middle-level managers with less clout to bring issues to the attention of top management.[45]

To a certain extent, lack of early assessment reflects classic locational theory: environmental and land-use regulations are generally less important than traditional locational factors in selecting industrial sites. This, however, does not mean that they can be ignored until late in the planning process, especially on major industrial projects. A company or project manager that does so risks unnecessary delays later. A General Accounting Office study concluded that there were few delays in the EIS process when the impact statement was done early, *before* a project proponent was irretrievably committed to a particular site, process, or plan.[46]

Two of the case studies in chapter 2 illustrate this point. In the Hampton Roads refinery case, HRECO spent little time studying the environmental ramifications of the site it chose. In fact, a government task force rated that site at the bottom of potential East Coast locations. Even if the Hampton Roads site were better than some others, however, HRECO might have avoided some of the opposition it encountered by identifying potential environmental problems earlier and presenting plans to mitigate them.

The Dow case teaches a similar lesson. Dow had sound business reasons for wanting to build a West Coast petrochemical refinery, but it gave only slight consideration to alternatives to the site it ultimately chose. Even then, environmental factors played no real role. Moreover, although Dow did anticipate some environmental problems at the site it selected (notably air pollution), its project team did not focus early enough on a host of other important environmental-quality issues such as oil spills and secondary-growth impacts that ultimately emerged as crucial issues. That failure caused delay, as regulators requested more information and opponents acquired ammunition for their battle.

The short shrift that corporations give to environmental issues in siting decisions reflects the fact that businesses also do their overall strategic planning—how much capital to invest and where—without giving much thought to environmental issues. A vice-president of AMAX, Inc., has said:

> A lot of business decisions are tactical and fail to link up with strategic decisions. Businesspeople making decisions in the environmental area need to start linking up environmental assessment and political assessment with a strategic business plan. It is surprising how many large corporations have business plans that say nothing about the external environment and the risks to the business of the possibility of [a project] being stopped.[47]

Many corporate executives and employees have a personal sense of environmental responsibility, but the dictates of corporate project planning often prevent an adequate and timely response to environmental and land-use issues. Project managers are rewarded if a facility is planned and built quickly. Few points are awarded if the project takes longer even though it is environmentally sounder and will cause fewer adverse impacts in the long term. Only if a project were ultimately blocked or seriously delayed would a project manager be likely to receive demerits.

In a book on environmental ethics in America, a leading observer of the environmental scene found that "the predominant ethic of business, centered around short-term results and a narrow identification of its interests, largely overlooks environmental concerns."[48] He was particularly disturbed to discover

> so few companies with an adequate institutional structure for environmental decisions, other than a unit charged with pollution control matters. In most large corporations today the chief executive officer's personal conviction for—or bias against—environmental responsibility sets the policy for the company. Rare is the company that includes in its corporate structure

a system that allows for the environmental impacts of all major decisions to be brought to the attention of top management where options can be presented for less environmentally harmful alternative solutions."[49]

Environmental Reconnaissance Statements

One way to ensure that environmental factors get evaluated early is for firms to require that each major capital proposal be accompanied by a reconnaissance statement that briefly outlines potential environmental and land-use difficulties. In addition to identifying potential hurdles, the statement could analyze how long it might take to secure approvals and how costly that process might be. It could also present strategies for avoiding difficulties, forecast the likelihood of the project proceeding smoothly through the regulatory process, and discuss contingency plans in case the project was blocked or seriously delayed. Such a statement need not be long—five to ten pages at most—to be effective; brevity would increase its chances of actually being read.

The prospect that top management would consider such a statement along with more traditional concerns could work a revolution in the ranks. Project proponents within a company would have to devote at least some time to checking out potential environmental constraints or risk losing face if their disclosure statement proved faulty.

In projects where the site was already chosen (a mine, say, and its attendant facilities), contact with government officials might be helpful at this point to sketch out proposals and test the regulatory waters. In addition, the reconnaissance report would allow project proponents within the company to adjust their proposals to avoid environmental problems later on, thereby reducing delay and costs.

By preparing a reconnaissance report, the proponents of a facility might find that their favored production technology would make it difficult to meet clean air standards in all but a few places. This would allow them to make changes before the firm was committed to a particular process. Or the analysis might reveal that there were likely to be large amounts of hazardous wastes to dispose of or that secondary-growth impacts would be significant; such information would be valuable later, after capital was allocated and the company began its search for a site or started negotiating with government agencies for permits. Several analysts have developed screens and checklists that help identify applicable regulations and potential pollution and land-use problems, depending on the size of a facility.[50] These aids can serve as an outline for the disclosure statement.

Who should prepare this environmental reconnaissance statement? Businesses that have already built similar facilities will have sufficient in-house expertise to evaluate potential environmental problems. However, only a handful of firms currently are analyzing potential regulatory hurdles facing major capital proposals. Mead Corporation, for example, has elevated its Human and Environmental Protection Department from the research level to being a major part of the company's overall planning; it now routinely reviews Mead's one- and five-year plans to ensure they meet regulatory requirements.[51] The department also analyzes individual projects, focusing on environmental issues as well as socioeconomic and secondary-growth impacts.

AMAX, Inc., operates in a similar manner. In 1973 it created a new corporate department, the Environmental Services Group, that reports directly to the chief executive officer. The group, which is now an AMAX subsidiary, reviews all long-range plans and each major capital appropriation request. If a potential environmental problem is discovered, the head of the department can lobby to have approval delayed until solutions are worked out.[52]

But what if the company is new to the game or small? In these instances, outside advice may be necessary. Fortunately, a growing cadre of specialists (many of whom at one time worked within the regulatory system) can identify important environmental and land-use considerations.

Once top management approves capital for a project and a list of potential locations is developed, the environmental reconnaissance statement should pay attention to particular environmental constraints that each site might have—for example, air-pollution problems or proximity to a critical natural area such as a wetland. Questions should also be raised about the capability of local governments to provide adequate infrastructure, and to handle secondary-growth impacts. State and local environmental and land-use regulatory requirements should be reviewed, not so much for detailed analysis or comparison with laws applicable to another site as for determination of potential project-stoppers.

Such a statement not only would identify potential hurdles and help to avoid locations that lack a fighting chance of success; it would also give project planners a better idea of how long building a project in a particular location might take and what costs could be involved in mitigating adverse environmental impacts. This information would also be invaluable in project planning later, after a specific site was chosen.

Improved Project Organization and Project Management

Just as governments need to improve their internal organization and use more-experienced regulators, so too must firms change the way they put together project teams.

The SOHIO case is a good illustration of how a project team should *not* be organized. Decision-making responsibility there was diffused, with no one on the project team in California having authority to make quick decisions. The problems caused by the failure of regulatory agencies to appoint specific contacts for projects were compounded by the lack of a corporate project head, which led to delays and frustration among regulators, particularly with regard to environmental impact analyses.

Project planners now appear to be learning from experience and to be drawing more frequently on highly trained personnel and consultants for help. But the learning process seems to be quite slow. Because relatively few pollution-intensive, controversial projects are built by any company over even a ten-year period, the payoff from these experiences has sometimes been a long time coming.

How should a company organize to meet the new environmental imperatives of the siting process? The major change needed is for site-search teams to have someone who can spot potential environmental and land-use problems and bring them to the company's attention. This person could be someone on the team itself, another employee of the company, or an outside consultant. If the final few sites under consideration for a project raise environmental or land-use questions, a more detailed analysis is prudent, as is consultation with those persons responsible for the engineering and design work that is probably under way. Unfortunately, few firms utilize such expertise in the site-search process, waiting instead until a site has been chosen.

Several corporations have compiled excellent records of responding to environmental and land-use criticisms once a site has been chosen. Usually, they assemble a separate project team to oversee detailed project engineering and design, the securing of government

RESPONSES TO THE REAL PROBLEMS 175

approvals, public relations, and the like. The composition and organization of this team can mean the difference between failure and success for a controversial project. One lawyer who has shepherded numerous big projects through the regulatory system has observed, "Management has turned the corner." He adds, half-facetiously, that he educated utility clients by "telling them they were really in the business of running pollution control devices whose by-product is electricity."[53]

The recent siting effort by AMAX, Inc., for a molybdenum mine in Gunnison County, Colorado, discussed in chapter 5, shows sound 1980s project planning.[54] Clear lines of authority were established so that government agencies could deal with one single, experienced project manager rather than a host of public relations people, scientists, engineers, and front-office types. The project manager could deal authoritatively with government agencies and make decisions quickly in negotiating sessions. Thus, if regulators wanted AMAX to spend additional funds on evaluating the impact of a particular facet of the project, the project manager could agree on the spot rather than having to seek approval from headquarters, which is time-consuming.

AMAX brought environmental specialists into the early stages of project planning to help pinpoint environmental and land-use issues so that the company would not be "ambushed" later in the permitting process. In addition, the company established a technical review committee to monitor and discuss technical aspects of the project on a continuing basis with the state of Colorado and the U.S. Forest Service.

One factor that has made such close contact with regulatory agencies work is that AMAX has taken pains not to compromise their independence. The state and local agencies have been cooperative but not overly friendly, unlike the Dow case, in which the local government appeared to be in Dow's back pocket. The impartial demeanor of the regulatory agencies in Colorado has helped persuade many local citizens and others interested in the project that environmental reviews will be balanced, not slanted.

Cooperation between the company, state and local regulators, and the local residents was also fostered by the fact that team members either had experience siting similar facilities in the state, so they were not novices in dealing with state agencies, or had worked in the area, so they were familiar with local concerns.

In contrast to AMAX, some industrial facilities have been delayed as project team members got on-the-job experience. In other instances, a firm lacking in-house expertise has sought local counsel. There is nothing wrong with that, of course, but one siting consultant has remarked that the most botched-up cases he has been called in to rescue have usually involved a company that retained a local attorney who happened to be a law school classmate of the company's general counsel rather than someone with environmental and regulatory expertise.

Early Public Involvement

Another key element that characterizes the better project planning efforts is their openness to the public. Many citizen-group suits are spawned early in the permitting process if companies avoid public involvement, causing people to feel that both corporate and governmental decisions are set in concrete. Later public hearings required by law seem to be window dressing.

There are several reasons for lack of early public involvement. The first, a throwback to early days of plant siting, is that many corporations assume that if a state or local government supports a project and promises to promote it, public approval is guaranteed. As the authors of the study of the BASF case observed (see chapter 1), perhaps one reason the company did not anticipate strong reaction against its project was that "it . . . assumed that support of a major governmental official would squelch any opposition that might arise."[55] In fact, such support may be a trap for the unwary. Unseemly coziness with a government agency may actually foster opposition.

Project planners also fail to involve the public early out of fear of scrutiny, worrying that opponents will become educated about project details and thus fight the development more effectively. Or businesspeople may fear that regulators will start asking more and more questions about the project, leading to demands for design changes. Notwithstanding these perceived drawbacks of early public involvement, firms that expect to site new plants in the 1980s will often be called on to disclose plans early, to meet with the public before plans are set, and to adjust plans to satisfy citizens as well as regulators. One recent study of energy facility siting counted these as advantages:

> Early public involvement, before plans have been finalized, has a greater

opportunity for modifying the application than would be the case after a final application has been filed and formal adjudicatory hearings have begun. In addition to defining issues, prehearing conferences may set schedules for the hearings, define how issues will be considered, and define discovery procedures.[56]

Similarly, a key element in AMAX's success was its early contact with governmental regulators and the public, including its agreement to disclose all company proposals before final decisions were made. Initial development concepts and background environmental data were shared with agencies and the public in a series of meetings and workshops before conceptual plans for the facility were completed. Company officials believe that by discussing their plans at this point—for example, by offering several alternative sites for a mill and transportation corridor—opposition to the project was defused. According to the AMAX project manager, "Our experience on other projects had been that we have always been able to gain reasonably good community support when we have done a good job of [environmental assessment] and have let the public know about it."[57]

The process will be unsettling to some businesspeople, however. One corporate executive who was involved in planning a major facility in the late 1970s explained:

> I would say that the thing that concerned us the most was the changing of the planning mode from the executive suite to the store front. This is a new experience and a typically conservative corporation doesn't reveal its plans. They like to surprise [competitors] and get some competitive advantage if it can be done! A firm typically does not expose the development process to the public eye.[58]

But his company took the plunge and opened up its planning process. "I guess this was more of a mental stumbling block than anything else when we finally recognized that this is the kind of thing that must be done to comply with the National Environmental Policy Act." The firm now feels that, as a result of open planning, it was able to expedite approval of the required EIS.[59] Other companies have organized citizen advisory committees to counsel them on mitigating project impacts and have opened "store-front" project planning offices that citizens can walk into for information about a project.

Still, many in industry remain skeptical. Some say they have built plants in, say, Louisiana without much worry about involving citizens beyond the statutory periods to comment on permits.[60] Others say that technical decisions should be left to the experts. Public hearings are seen as inefficient forums that drag out the process while producing irrelevant information.

In some areas of the nation the old way of building plants will survive, at least for the moment, but there is evidence that those areas are becoming fewer. Those who thought development in the energy-rich West would be easy, given the antiregulatory attitude of many citizens there, have been surprised. In Louisiana, which has been thought "soft" on environmental pollution, increasingly frequent problems with contaminated water supplies and other pollution incidents have alerted citizens to the negative aspects of industrial growth.[61]

Even if early citizen participation in the siting process does not guarantee faster approval, it does decrease the chances that project proponents will commit themselves to a site that will later engender great controversy and that the project will be blocked altogether.[62] Also important, an open planning process allows other reforms such as consolidated hearings and expedited decision-making schedules, since most people will agree to allow speedups if they have confidence that their interests will be protected.

Greater Attention to Indirect, Quality-of-Life Impacts

Although firms do seem to be doing a better job of evaluating direct project impacts, such as air and water pollution, they have been far less sophisticated in dealing with indirect impacts. Dow, for example, was slow to recognize and respond to concerns about industrial growth and residential development that the project might spur and to concerns about the potential impact on the supply of agricultural land and the overall quality of life. This will be an important next step in the learning process for project planners.

A recent survey of environmental/industrial conflicts across the United States from 1970 to 1978 concluded that the scope of such conflict is broadening. "The issues at stake are changing, with land use, social impact, and human health concerns rapidly on the rise as central matters in contention. The name of the game is no longer simply ecology, but rather the overall quality of human life."[63]

Support can be found for such a conclusion, particularly where a new facility is being built in a previously undeveloped area (known in industry as a greenfield or grassroots site). It was certainly true in the Dow case and can be seen with increasing frequency in mineral- and energy-facility siting cases in the West, where industrial development is a new phenomenon in many areas. One of the highlights of AMAX's planning effort in Colorado, and elsewhere, has been its sensitivity to a project's secondary-growth impacts.

A former vice-president of U.S. Steel, however, may have expressed the prevailing attitude of corporate executives. He complained that even though U.S. Steel had conducted what he thought was a very open planning process in its efforts to site a major steel mill at Conneaut, Ohio, the company ended up in court because opponents worried that the facility would harm agricultural land in the area and displace jobs in Youngstown and Pittsburgh. "If these secondary, socioeconomic impacts are elevated to equal level with environmental impacts as a result of this case, another dimension will be added to corporate planning."[64]

Cases from across the United States make clear that this dimension has already been added. According to a local county commissioner from Rio Blanco County, Colorado, which is facing major energy-related developments: "We would give up our good life style because these projects are necessary, *but* we won't pay for the privilege of doing so. The vast majority of mayors and town councils feel like I do. We owe a duty to our constituents and want industry to know the real cost of doing business."[65] One industry project planner gave this example:

> A few years ago, many rural Colorado county commissioners were uncomfortable with imposing even the most basic land-use controls. As an illustration of how far things have gone in the past few years, the Mesa county draft siting regulations [require] . . . a full assessment of the impact of a proposed energy facility . . . mechanisms for dealing with all direct and indirect costs . . . [so that] the pace of growth caused by energy development not exceed the capacity to mitigate and absorb adverse environmental, economic and social impacts of such growth . . . and the cost of growth impacts . . . shall not be borne by local residents and industries.[66]

He concluded that "rural local government can hardly be considered a bystander in the regulatory process today."

In fact, many local communities now focus on land-use and overall growth issues, while relying on state agencies to handle technical environmental issues that require more expertise. In the West, energy and mining companies are finding that the key to local approval often is the development of a mitigation package that ensures that local quality of life will not be sacrificed to a national search for natural resources.[67] For example, one energy-project proponent, Western Fuels-Utah, Inc., had to sign a socioeconomic mitigation agreement to get the approval of Rio Blanco, Colorado, and other local governmental units. The company will spend the money on project reviews, monitoring, and infrastructure improvements.

Attention to secondary-growth impacts also makes a great deal of sense to any industry that needs a stable work force and that cares about the quality of life its employees will have. Utilities, for example, have found that work-force problems are costly. A former Westinghouse official recently attributed a nearly 100 percent worker-turnover rate (and consequently half of a $100 million cost overrun) at three 1,500-megawatt power plants in Rock Springs, Wyoming, to inadequate emphasis on mitigating secondary-growth impacts.[68] But he also saw signs of positive response among several progressive utility companies. Similarly, clean, high-technology companies have been discovering that, unless they participate in local planning efforts and help mitigate adverse impacts that their plants engender— more traffic, crowding, air pollution from automobiles, escalating housing costs—they may find it difficult to attract upper-level employees who place great weight on an area's quality of life when selecting an employer.

Adoption of New Ways to Avoid or Settle Disputes

What about the prospect of being taken to court even after permits are issued? After all, it only takes one disgruntled person to sue.

The prospect of litigation and lengthy judicial review has been the subject of several reform efforts, most of which would place strict limits on the use of judicial forums to settle cases. After Dow killed its project in California, several state legislators proposed a law requiring project opponents to post a bond equal to 5 percent of the project's value. In the Dow case, a bond of $25,000,000 would have been required—justice by tollgate, critics claimed.

Another commonly suggested approach is to set up special review courts. But these courts have a mixed record at best. For example, the Temporary Emergency Court of Appeals (TECA) that would have heard cases under the EMB law is a prime example of good intentions gone awry. TECA cases are presided over by three-judge panels whose justices rotate into Washington, D.C., from other federal circuits. Cases that come to TECA are thus handled by many different judges. The result, according to lawyers who practice before TECA regularly, is that decisions are often poorly reasoned, because judges are not particularly familiar with the specialized law they must interpret. Even worse, because so many judges hear cases, the results are often contradictory.

Allowing direct appeals to a state supreme court raises other questions. Why should review of an industrial-facility permit be put ahead of criminal cases on a high court's docket? Which is more important? Moreover, is it good policy for state supreme courts to be hearing cases where legal and factual issues have not been narrowed by lower court decisions? Finally, what if a state supreme court fails to expedite a case? Experience in several states indicates that if major projects are granted expedited review, the result is often more delay, not less. The SOHIO case stands as testimony to the pitfalls of special treatment for plant-siting cases.

There are also proposals to cut off judicial review if appeals are not filed quickly after a final administrative decision. Experienced environmental lawyers assert that such a provision would only lead to quick filing of appeals with little forethought. Thus, rather than have time to consider the prospects of success in a case, project opponents might be backed into a corner and forced to sue.

A business school professor who has studied the siting process has further observed that the judicial system in this country currently serves a positive function as a safety valve in siting disputes. He attributes the relative lack of violence and mass demonstrations in U.S. environmental disputes—compared with Europe, that is—to the availability of the courts for redress of grievances.[69]

But what should be done about the specter of lengthy judicial reviews that surfaces in many siting battles? A partial solution lies in the way project proponents and regulators approach potential conflicts. The case studies show that the best insurance is to follow carefully all legal requirements in order to insulate an agency decision from challenge. In some cases, particularly at the local level, this means that industry will need to bear the burden of helping regulatory agencies satisfy the law. In the Dow case, rather than push Solano County for quick approval, the company would have been far better off carefully advising local officials on proper procedures for rezoning the site from agricultural to industrial and for cancelling the open space preservation contract on Dow's property.

Care in complying with substantive and procedural mandates could avoid lawsuits in some instances and help ensure ultimate success if an action is filed. There are, however, other promising approaches that can help defuse siting battles. The greatest hope of addressing such problems lies in adopting new approaches to avoiding and settling disputes—especially, mediation and mitigation.

Mediation

Mediation is defined by one of its leading practitioners, Gerald Cormick of the Office of Environmental Mediation, Institute for Environmental Studies, in Seattle, Washington, as a voluntary process in which disputing parties together explore issues dividing them.[70] A mediated settlement is reached when all parties agree they have achieved a resolution that is technically, financially, and politically workable. The mediator, who acts somewhat like an arbitrator in a labor dispute, generally has no power to force a settlement but rather acts to define the framework within which the parties will negotiate, to oversee substantive negotiations, and to help implement a settlement.

Cormick stresses that mediation will not work in all cases, but works best when the following conditions are satisfied:

- Mediation is most appropriate at the point in a dispute when the issues have been defined, the parties are visible and highly involved, and there is some sense of urgency to resolve the conflict.
- Mediation requires the existence of a relative balance of power among the parties. The ability of each party to exercise some sanction over the other is necessary to ensure that all parties will enter into the negotiation process with "good faith."
- Mediation results in compromises being made. Therefore, certain issues of principle or law are not appropriate for mediation.
- Mediation is appropriate in situations where there are reasonable assurances that responsible authorities will implement an agreement reached by the disputing parties.
- Mediation requires the participation of a mediator who operates from an impartial base and whose primary role is to promote agreement among the conflicting parties.[71]

The use of mediation is not yet widespread and has been most prevalent in cases where government agencies were the developers. But results over the last few years have been impressive.[72] In June 1981, 12 negotiators representing environmental, economic, and local government interests signed an agreement to guide water-related developments along the Oregon side of the Columbia River, heading off what had the makings of an acrimonious and time-consuming dispute.[73] Mediation also helped in what was perhaps the longest-running environmental controversy in the nation—Consolidated Edison's pumped-storage power plant near Storm King Mountain in New York.

Mitigation

Mitigation, already discussed as a tool for dealing with negative secondary-growth impacts, also is a promising method to resolve disputes nonjudicially by providing compensation, financial and otherwise, for the potential adverse environmental effects of projects.[74] As with mediation, the use of mitigation by industries is not widespread, although it was routinely used during the 1970s by sophisticated residential developers.

Recently, however, some industrial developers have begun to use this technique. In Wyoming and Nebraska, for example, the project developer of the Grayrocks Dam won approval in the late 1970s from environmentalists by establishing a trust fund to protect whooping cranes threatened by the project.[75] And in the San Bruno Mountains near San Francisco, a large residential development that would have been killed under a prior law has been able to proceed because the developer formulated a plan to preserve a rare butterfly.[76] Without that plan, the outlook for the butterfly's survival was dim, even if the project had been rejected.

When will compensation work best? According to one leading proponent, the following characteristics describe the ideal situation:[77]

- Few parties to the dispute should exist. Negotiations with several parties can be very complicated.
- Opponents should be geographically defined and easily identifiable, to enable tailoring compensation to a discrete area.
- Opponents should be well organized, making negotiations with them easier.
- Outcomes that are mutually acceptable should exist. Incentives will fail if one party adamantly opposes a project.
- A project's impacts should be clearly traceable. Establishing a causal link makes it easier to define compensation.
- Re-creating the status quo should be possible. Unless some impacts can be adequately mitigated, opponents will be unwilling to negotiate fully.
- Parties should offer a binding commitment. If one party appears unable to abide by any agreement, negotiations are doubtful.
- Hostility should be absent, since it can poison any negotiations.

Like most attempts to improve the siting process, the idea of compensation contains pitfalls. To some people, it smacks of bribery, and, in fact, it may be little more than that in some instances. (Chapter 4 details the problems a German utility experienced when it appeased opposition to one of its power plants with an extremely generous "compensation" package for a nearby community and local opponents.) Some legal hurdles also exist. In some states, the legality of a community extracting a compensation package from a project proponent is highly questionable at best, particularly where the compensation is not directly related to the project's impact. However, mitigation as a tool to ease disputes has been codified in the Federal Endangered Species Act.[78] As discussed earlier, the act allows incidental effects on the habitat of endangered species, provided a project proponent develops a conservation plan to ensure its survival.

Mediation and mitigation do have their limitations, but they should still be of great value to industries in future siting efforts. This may be particularly true at the local level, where state and federal grants that might be used to cope with the direct and indirect impacts of new development are being cut back.

CLOSING THE COMMUNICATION GAP

Poor communication between industry and regulators is a major problem. One of industry's primary complaints about the regulatory system is that regulators have seemingly insatiable appetites for more information. In response, regulators say they must have adequate information to make decisions and that project proponents slow things down by submitting incomplete applications or overlooking key environmental issues. A 1982 federal EPA analysis of delays in granting permits for major energy projects found that of ten proposals that were seriously delayed, six had been held up because of incomplete or inadequate information in the applications.[79] Although industrial firms sometimes purposely withhold information for proprietary reasons, more often the problem can be traced to poor preparation.

The communication gap has another side: some major projects have been needlessly delayed because proponents could not secure information about what permits were required, from whom, and when. In other instances, corporate planners have secured the informal blessing of government agencies to proceed with a project only to learn that they and the agencies failed to identify potential

RESPONSES TO THE REAL PROBLEMS 185

environmental problems (perhaps the habitat of an endangered species protected by federal law) or steered the facility to a site that, on further study, did not really have a fighting chance of being approved.

Closing the communication gap is an essential element in making the regulatory system work better. Both government and industry have roles to play here.

Adequate Information for Regulators

For a variety of reasons, regulators do not always have adequate or timely information on which to base their actions. As one industry project manager observed, "Like our own engineers and managers, government managers need information as early as possible so that they can understand how the project affects them and be prepared to make quality decisions with the least amount of time and effort."[80] This shortcoming often leads not only to delay in processing permits as regulators seek more information but also to poor environmental assessment of a project. What causes this problem?

Meshing the Regulatory Process with Project Planning

The fact that the regulatory process does not mesh well with the corporate project planning process presents a dilemma: how can regulators be satisfied that they have sufficient information to indicate their preliminary approval of a project before a firm spends large sums on design and engineering?

There are no easy solutions to this particular quandary. Until regulators are convinced that a project has been evaluated adequately, thereby protecting them from charges that they have failed to exercise due care, and until they feel that enforcement will be effective and expeditious, they will continue to demand as much information as possible. Moreover, lack of information is likely to stoke environmental opposition to projects. There are, however, several steps that can reduce this problem.

One, already taken by many agencies, is use of preapplication meetings between company officials and regulators. Such meetings can help companies identify both the permits they will need and the information that regulators expect to see in an application. Regulators can also bring potential problems to the firm's attention. Some states have special permit offices to facilitate preapplication meetings with state agencies.

Preapplication meetings can only accomplish so much, however, as witnessed by the Dow and SOHIO cases. What firms really want is an indication of whether a project will ultimately receive approval before they provide regulators with detailed data. In contrast, government regulators usually do not like their positions (in the words of one high-ranking former federal official) "smoked-out" early.[81] And, of course, they may be unable to reach a decision until they have detailed information on which to base it.

A workable middle ground might be to issue what in the land-use regulatory field is called an outline or concept approval. When a large developer approaches a local government with plans for a phased housing project, outline approval is often given covering the major facets of the development like density, roads, and community facilities. If the developer agrees to conditions regarding the major elements, then assurance is given that final approval will be granted later under normal circumstances. Details must still be worked out, but unless changes are proposed by the developer, the local government agrees not to reject the proposal. Outline approval can guard a developer against spending thousands of dollars in design and engineering work, only to be turned down later.

The use of outline approvals may be trickier when dealing with permits involving air or water pollution, particularly where there might be a direct impact on human health. For example, new evidence might emerge that requires tightening of controls to protect human health. That problem could be dealt with by a proviso forbidding change unless the regulatory agency could prove that public health was endangered. With a preliminary approval in hand, project proponents should be far more willing to develop detailed design and engineering information and share it with regulators. Negotiations over details would still take place, but in the context of ground rules set out in the preliminary approval.

Outline approvals would be valuable in major multistaged projects involving a series of production processes, such as the Dow petrochemical complex. Other candidates for such approvals are plants or processes with which regulators already have a good deal of experience. A special EPA task-force report on industrial project permitting has encouraged the use of what it calls "conditional" permits for familiar projects: "The early-approved permit still has conditions such as final approval of detailed specification. But the plus is that the applicant can demonstrate, early on by analogy, that what he plans to do can meet environmental requirements. This approach can save both applicant and reviewer time and effort."[82]

The EPA report notes that the state of Oregon has already used conditional permits for some projects and that several EPA regions are experimenting with the idea. As regulators gain more experience in dealing with large facilities, conditional permits may become more attractive. Although the report notes that relying on existing processes to speed approval may discourage development of new control technologies, it concludes that "ambient standards with increasingly strict performance standards . . . and similar requirements would appear sufficient to overcome this potential."[83]

Trade Secrets and Land Speculation

The communication gap can also be traced to fears that a company may inadvertently disclose important trade secrets or marketing information during the permitting process or that, once the location of a project is disclosed, speculators will rush in and drive land prices sky high. The basic clash in both instances is between a company's legitimate need for confidentiality and an open regulatory process in which agency officials and citizens have access to such information as plant location, pollutants, and production processes so that they can make an informed evaluation of a project.

With regard to trade secrets, two important points should be kept in mind. First, existing laws, particularly the Freedom of Information Act (FOIA), do protect trade secrets from disclosure; rarely do agencies reveal them. However, industry typically demands that any proprietary, commercially valuable information be protected, and that is where the conflict really takes place.[84] Second, most attempts to force disclosure of sensitive commercial information are carried out by representatives of other firms seeking to gain a competitive advantage, not by environmental or citizen groups opposing a project. In 1980, for example, 85 percent of more than 33,000 FOIA requests filed with the Food and Drug Administration were from companies seeking information about their competitors. (These two points are not offered as an argument for or against disclosure, only as a perspective on the scope of the problem.)

Industry does appear to have legitimate fears, but those concerns must be balanced with regulators' and the public's need for information. A good illustration of how concern about trade secrets can be taken too far comes from the state of Virginia. In 1976, after the James River was contaminated with Kepone, a new state law was enacted requiring manufacturers to report in detail on any dangerous substances they produce. However, because the law makes state officials who disclose trade secrets subject to triple monetary damages in a civil suit, as well as to criminal penalties, state employees are understandably hesitant to delve into such reports.[85] The state legislature is now trying to find a reasonable compromise.

One approach that may strike an appropriate balance, albeit one not completely acceptable to either environmentalists or industry, is the so-called reading room option.[86] Regulators would have broad access to information needed to evaluate a project, but access by the public and competitors would be somewhat restricted. Data and other information about a project would be provided in an agency reading room, and copies could be made only by specified persons. Excluded would be representatives of any competing firm. Restrictions would also be imposed on disposition of any copies made. Because of potential enforcement difficulties, companies might be given the right to pursue violators on their own.

Another useful tack was first suggested in the above-mentioned EPA permitting report: private rulings similar to those of the Internal Revenue Service (IRS) or the Anti-Trust Division of the Justice Department.[87] A firm might wonder whether a new production process could meet air-emission standards, for example. Instead of dickering with regulators over how much information about the process was needed before a permit application could be considered complete, the company might approach the EPA for a private ruling in which it would disclose more information. Like IRS rulings, the EPA ruling would be made available to the public, with the facts of the particular case disguised to avoid giving away trade secrets or marketing information. Also, the ruling would be binding on the EPA.

Another knotty problem can be land speculation. Typically, companies like to buy land for developments quietly, often through a third party, to keep prices down. If an industry embraces an open planning process and reveals several alternative sites for a project, it runs the risk of having speculators buy land that will ultimately have to be purchased at an inflated price by the company.

Although the answer would seem to be the use of long-term options on several parcels until the regulatory process proceeds far enough to settle on one site, industry representatives say landowners are reluctant to enter into such agreements and that these long-term options are very expensive. However, experience with large-scale residential and commercial projects argues otherwise. In the early 1970s, smart developers began entering into long-term land options contingent on local governments' granting land-use approvals. That practice is now fairly commonplace, as developers and landowners realize that land-use and regulatory approvals are less automatic than they used to be.

This solution is not cheap, but it is far better than buying land outright (as Dow did in California) and risking disapproval. Indeed, when viewed against the possibility of having a project blocked by citizen opposition that essentially had doomed a chosen site from the start, this approach can seem quite economical.

Scoping

In addition to these adjustments, government agencies can take another particularly useful step to improve the amount and quality of information available to them: the use of "scoping," a technique developed by the CEQ as part of the EIS process. In scoping, all agencies involved in evaluating a project get together and identify significant issues in the application process. As the architect of the CEQ provisions has written, the goal is to ensure that "the truly significant issues will be given adequate study, while those that are not significant will not need to be analyzed at length."[88]

By focusing resources on key issues, agencies have been able to develop more in-depth information on which to base decisions and have avoided holding up proceedings in later stages to gather more data. Several states have adopted similar procedures. In California, the Office of Permit Assistance helps organize the scoping meetings. In Colorado, the joint review process office assumes that role.

The results to date at both the federal and state levels are very encouraging. For example, early and comprehensive identification of significant environmental issues involving the proposed Alpetco refinery at Valdez, Alaska, helped the EPA and other agencies coordinate environmental documentation needs and data collection. Information needed for the federal EIS was also used for air- and water-quality permits. Together with other procedural innovations, this initiative helped provide an efficient and expeditious review of that proposal.[89]

190 PART III: RESPONSES TO THE NEW REALITIES

Helping Local Governments

Knowing what questions to ask industry usually is not a problem for federal and state regulators. Increasingly, however, local governments also face important regulatory responsibilities when big projects are proposed. This is due to the continued movement of industry, for a variety of reasons, to less densely populated or developed areas. In the Dow case, for example, a small county with limited environmental expertise took the lead in evaluating the company's petrochemical complex. In the West, small rural counties have been entertaining proposals for massive energy projects.

Size is not necessarily an indicator of how well local governments will respond. Sparsely populated Rio Blanco County, in western Colorado, recently negotiated a lengthy agreement with an energy project developer that deals with socioeconomic issues in a very comprehensive manner [90] Overall, however, the record is mixed at best, particularly with regard to evaluating nonpollution environmental issues such as land-use planning, wildlife protection, and more general secondary-growth impacts.

Federal and state agencies can generally be relied on to evaluate air- and water-pollution impacts, but for the most part small local governments find themselves on unchartered waters when it comes to major industrial projects. Some may, in fact, only confront such issues once or twice in a decade. The result: projects are approved without adequate information or review in some cases, or, increasingly, they are turned down because local government officials do not feel confident they can deal with secondary-growth and land-use impacts.

How can local governments get the information they need, evaluate it properly, and act to cope successfully with industrially induced growth? In part, the answer is to provide them with adequate money and resources to plan sensibly for new growth, to help them identify needed infrastructure improvements, and to assist them in preparing local institutions like hospitals and schools for a surge in demand.

Who is responsible for such aid? Senator Gary Hart of Colorado, among others, has been a strong proponent of federal impact assistance for energy boom towns associated with developments on federal land.[91] The rationale is that since the federal government will make a good deal of money in these cases through leasing fees, a small amount should be returned to localities affected by the growth such developments precipitate. But strong arguments can be made that the federal government already gives states a large share of federal leasing revenues (for example, in 1979 BLM gave states that contain large amounts of federally owned land $208 million pursuant to the 1920 Mineral Lands Leasing Act[92]) and that growth management is a state and local, not federal, responsibility. This position—that the federal government is already providing enough help—carries considerable weight on cost-conscious Capitol Hill.

If direct federal aid is not forthcoming, local governments will look first to the states or federal permitting agencies to help manage secondary growth. To date, both the states and the federal agencies have checkered records. Although several states have enacted legislation to provide "front-end" financing for local governments, allowing them to collect projected property tax revenues from developments early or to borrow against such projected income, these programs are hobbled by a number of practical and legal problems.[93] Utah has earmarked 32.5 percent of federal funds received from BLM under the Mineral Lands Leasing Act for a Community Development Account that is used to assist localities affected by natural resource development.[94] As yet, however, none of these schemes has proved to be a complete answer.

In this context, the most effective approach to local secondary-growth problems may be to establish state entities that can give local governments direct assistance in assessing resource-development impacts. The Wyoming Industrial Council has proved adroit at, first, helping local governments assess secondary-growth problems and, then, imposing conditions on state permits to address them.[95] Several other states have less ambitious alternative programs for aiding local governments faced with major projects. They have established teams of experts who are on call to help local governments plan for such growth.

Federal agencies, such as the EPA, the Army Corps of Engineers, and the BLM, will be called on to play a similar role in development on public lands or for projects requiring a federal impact statement. What results can be expected?

In the past, the BLM has been relatively adept at assessing the impact of resource developments on the public lands themselves, but local governments and environmentalists complain its analysis of effects on nearby communities has been inadequate. In the late 1970s, the Office of Special Projects (OSP) in the BLM began to improve that record; as noted, however, OSP has since been abolished.[96]

Federal agencies such as the EPA started similar initiatives in the late 1970s, but budgetary cutbacks and an antiregulatory mood do not augur well for continuing or improving such efforts. In Colorado, the state legislature recently cut funding for a state "circuit rider" program that sent land-use planners to local governments that requested their advice. The program was axed even though local governments footed part of the bill for such advice and the program was popular in high-growth areas.

If federal and state agencies fail to lend a hand, then the burden will most likely fall on industry, either from choice or necessity. Local governments will increasingly demand impact mitigation assistance from major industrial projects, as well as money to hire outside consultants for project review. This has already begun to occur. For example, the Rio Blanco, Colorado, mitigation agreement mentioned above will reportedly cost Western Fuels-Utah $100 million. There should be little doubt that this practice will become common, even though local governments in some areas of the country still seem not to care much about land-use, environmental, and socioeconomic impacts.

Farsighted firms, rather than waiting for local governments to demand aid, have decided that volunteering assistance early will pay dividends later and make project approval more likely. Clearly, environmentalists will increasingly demand, as direct responsibility for federal environmental laws is devolved to state and local governments (for example, as the Clean Air Act and the Surface Mining Control and Reclamation Act stipulate), that adequate resources be made available for sound reviews. Failing that, more cases will almost surely end up in court.

More Information for Project Proponents

There is another side to the communication gap. Although a project proponent must bear some of the blame for inadequate, ill-timed environmental assessment, its task is made more difficult by shortcomings in the regulatory system.

Perhaps most noticeable in the Dow, SOHIO, and Hampton Roads cases (and other siting disputes that occurred during the past decade) was the inability of government agencies to provide corporate project planners with sufficient environmental quality data or resource surveys—for example, the location of critical wetlands or ecological sites or information about regional air quality—before a site was

RESPONSES TO THE REAL PROBLEMS   193

selected. As a result, firms must often take precious time to develop necessary data (particularly concerning air quality in many areas of the country) if they are to avoid choosing a site that ends up having serious problems. In other instances, like the SOHIO case, different agencies may try to steer a plant to different sites based on their own resource-protection agendas. With more information, corporations would neither waste time early in the permitting process nor be attracted or steered toward sites with little chance for approval.

Without short-circuiting the system, government agencies can act to improve the quality of corporate decision-making and planning while speeding ultimate review. In fact, several jurisdictions and agencies have begun aggressive efforts to bridge the communication gap. Two particularly promising approaches are identifying off-limit and acceptable sites for development and compiling permit registers and guides.

Identifying Off-Limit and Acceptable Sites

In some of the most interesting efforts to bridge the communication gap, public agencies either identify sites that have important environmental values (that, for all intents and purposes, are off limits to developers) or specify those with a "fighting chance" to receive development approval.[97]

For example, federal agencies have recently completed detailed resource surveys of the Atlantic Coast that should be very valuable in project planning.[98] These surveys were prompted in part by the HRECO/Portsmouth refinery dispute (see chapter 2). The Office of Coastal Zone Management (within the National Oceanic and Atmospheric Administration) has prepared a coastal atlas that identifies areas of critical environmental concern, such as oyster beds, and rates them according to importance and sensitivity. In a complementary project, the Fish and Wildlife Service has produced a series of maps to show fish and wildlife resources and habitats along the East Coast. Other federal agencies such as the BLM and the U.S. Forest Service have been busy surveying their own lands to identify critical environmental areas and historic and archaeological resources. These inventories, while far from complete, should help industry avoid stumbling into an environmental hornets' nest when planning projects (like mines and energy developments) on public lands.

The state of Michigan currently has an ambitious program to survey and inventory the state's land base.[99] Pursuant to a state resources inventory act enacted in 1980, the Michigan Department of Natural Resources has begun a high-quality aerial survey program of land uses across the state. The information is double-checked by a ground survey and then fed into a computer. Local government and industry officials in the state are enthusiastic about the program's potential, although funding cutbacks may slow its implementation. Industry will be able to use the information in considering suitable development sites, and local governments will receive valuable assistance in making sound land-use decisions, such as protecting prime farmland while allowing development of less important parcels.

Having a state agency identify sensitive environmental areas can provoke local opposition, particularly in the West, out of fear that it is the first step toward state control. But from an industry perspective, the information can be critical. Each state will have to solve this often thorny issue on its own.

Perhaps the most interesting and successful survey of natural resources is one initiated by the Nature Conservancy, a private non-profit organization that stores and manages information on natural ecological diversity.[100] In this effort, fieldworkers gather data about rare plant and animal species, various types of native plant communities in a state, and aquatic systems. As of 1980, 23 states and the Tennessee Valley Authority had adopted systems patterned after this Nature Conservancy program. Because of computerization and frequent updating, the information in this type of inventory is both current and readily accessible. In addition, it indicates how often specific species and ecological communities occur, so attention can be focused on those that are declining. Firms planning new facilities would be wise to consult such inventories to avoid potentially controversial sites.

The Association of Bay Area Governments (ABAG) in the San Francisco region has started a similar program to identify sites for industrial development.[101] Although it failed in earlier efforts to *select* appropriate sites for industry, ABAG still compiles information on property in the region that is zoned for industrial development. The group also identifies potential air- and water-pollution problems and catalogs other important site characteristics such as water supply. All this information is fed into a computer and made available to industries seeking to locate in the area. A company does not have to select one of these sites, but it knows that it will probably have fewer environmental headaches if it does so.

RESPONSES TO THE REAL PROBLEMS   195

A variation on the ABAG site-resource survey approach is under way in Grays Harbor, Washington.[102] Grays Harbor, an estuary on the Pacific Coast, is an important port and industrial center and supports major commercial fishing. It is also one of the most ecologically valuable estuaries on the West Coast, providing food, shelter, and nursery areas for a wide range of wildlife and marine animals. Longstanding controversy over the development of Grays Harbor intensified in the mid-1970s when plans for several major industrial projects were announced. In an attempt to resolve potential conflicts, an Estuary Planning Task Force, with representatives of various government agencies, was assembled to develop necessary technical information and then devise a comprehensive plan on where development should and should not take place. After much deliberation and compromise, agreement was reached on areas suitable for development as well as those of critical biological importance.

This planning process has been criticized for not including all interested parties (no environmentalists or industry representatives were on the Grays Harbor task force) and for committing agencies to particular sites rather than simply identifying suitable ones. Still, Grays Harbor's effort is worth examining as a way to provide industry with more information and greater predictability in locating new plants.

The idea of identifying acceptable sites for industry is also embodied in federal coastal zone legislation.[103] Before state coastal zone management plans can be approved, they must identify sites with unique geologic or topographic importance to industrial development—for example, a deepwater harbor—and ensure that facilities of greater than local concern are not arbitrarily or unreasonably excluded or restricted.

These resource surveys hold a great deal of promise, but recent cutbacks, particularly at the federal level, severely reduce their potential.[104] When budgets are cut, resource surveys and monitoring programs are usually the first activities to go. Thus, the ABAG program has slowed markedly since 1979, the Michigan program is severely underfunded, and many federal agencies have abandoned survey work or curtailed it severely.

Some claim that the federal cutbacks are based on the idea that if you don't know what's there, you don't have to worry about it. Whatever the motive, the cutbacks are penny-wise and pound-foolish. A few projects may get approved without adequate surveys or sufficient environmental impact data, but more are likely to be delayed as agencies are forced by statutory mandates and lawsuits to collect information before issuing permits. In a recent battle over energy development in western Wyoming, for example, environmentalists who were upset over the absence of good background resource data for permit review threatened to tie the project up in court. A fight was avoided only when industry agreed to foot the bill—$1 million—for such work.[105]

Compiling Permit Registers and Guides

In addition to providing valuable information about specific sites, government agencies must tell project proponents what permits they need, as well as when and how to get them. Until recently, only a few states and local governments had taken the relatively easy step of compiling permit registers to advise industry. Colorado and Florida, for example, have published useful "how-to" books for developers on state environmental permits.[106] And, as already mentioned, several states have taken the idea a step further by establishing industrial escort services.

Several local jurisdictions have also put together guides to the permit process that explain what permits are needed and that list the names and telephone numbers of the people who can provide them.[107] In the state of Washington, for example, 38 counties were operating some type of information center in 1979 to provide permit details to developers. During the Carter Administration, CEQ, under contract with the Environmental Law Institute (ELI), began developing a guide to federal environmental permits. A draft was produced by ELI but was never published by CEQ.[108]

The problem with most permit manuals, however, is that they become dated as soon as laws are amended or new ones enacted. Georgia has tackled this problem by putting its permit requirements on computer.[109] All a developer need do is visit the state DIT and tell officials the type of project being planned. The computer produces a list of needed permits and an explanation of each one.

Most of the innovative programs under way within government agencies will close the communication gap: expert project teams, the scoping process, and informal preapplication meetings. All these initiatives get more information to project planners, thereby making their lives easier without circumventing environmental reviews. If industry learns from the past and meets its challenge to adapt with

RESPONSES TO THE REAL PROBLEMS 197

equally innovative thinking, the industrial siting game will be played more smoothly.

REFERENCES

1. For a useful compilation of state streamlining efforts, with names of agency contacts, see Temple, Barker, and Sloane, Inc., *Streamlining the Permitting Process: A Survey of State Reforms* (Washington, D.C.: U.S. EPA, Office of Policy and Resource Management, 1982).
2. Final regulations were issued by CEQ on December 29, 1978, and can be found at 43 Fed. Reg. 55978.
3. Interview with Nicholas Yost, former CEQ general counsel, in Washington, D.C., February 1980.
4. 40 CFR Part 1501.5-.6.
5. Ibid.
6. Nicholas Yost, "Environmental Regulation—Myths, Realities and Reform," *The Environmental Professional*, vol. 1 (1979), p. 258.
7. Details on the OSP are from an interview with David C. Williams, former OSP chief, Washington, D.C., February 20, 1981, and from discussions at various later dates in 1982 with him and with Frank Gregg, former BLM director. Interior Secretary James Watt later dismantled OSP. The environmental impact statement team feature of OSP, discussed later, has been retained, but Williams asserts that BLM has lost the broader view of energy projects it should be taking as well as the coordinating function OSP played. According to Williams, OSP was killed because it was called "special" and was a Carter Administration initiative. Other officials add that OSP was viewed with suspicion by state BLM directors who resented having to share authority over big projects with a Washington-based office.
8. U.S. Bureau of Land Management, *A Review of the Bureau of Land Management's Energy Facility Permitting Process* (Washington, D.C.: BLM, 1981).
9. Williams interview.
10. This information is taken from interviews with Sidney Galler, U.S. Department of Commerce, March 1980, and conversations with J. Robert Ferguson, Jr., former senior vice-president and assistant to the president of U.S. Steel Corporation, June 1979.; see also Stanley Margolin, "The Two-Minute Mile," in *Early Corporate Environmental Assessment* (Washington, D.C.: U.S. Department of Commerce, 1980), p. 17.
11. Some opponents claimed the EIS was inadequate because it did not evaluate the impact of the project on other steel-producing areas of the country in terms of employment, nor did it look at the national need for such a project. For a critical discussion, see Tri-State Conference on the Impact of Steel, "Environmental Impact of New Industrial Plants, The Case of Conneaut," *Utah Law Review* (1980, no. 2), p. 331.
12. Interview with Wade Hopping, of Hopping and Boyd, in Tallahassee, Florida, March 26, 1980.
13. Interview with Harris Sherman, of the Denver office of Arnold and Porter, November 12, 1981.
14. See U.S. Bureau of Land Management, *Review of Energy Facility Permitting Process*, p. III-5, for a discussion of how lack of agency project managers has

198 PART III: RESPONSES TO THE NEW REALITIES

hampered BLM energy-facility permitting.

15. This description is based on interviews with Robert Rotan, John Gilliland, and Ronald Mayhew of the Georgia Department of Industry and Trade and with J. Leonard Ledbetter, director of the Georgia Environmental Protection Division, in Atlanta, Georgia, March 5-6, 1980.

16. This information is based on interviews with Dean Misczynski and Laurie Wright, California Governor's Office of Planning and Research, Sacramento, California, February 22, 1979, and later discussions during 1981 and 1982 with Deni Greene, director of OPA. See also Barry Steiner, *Mediation to Alleviate Conflict—The California Permit Streamlining Act*, presented at an American Law Institute-American Bar Association course on Environmental Regulation of New Land Development in the 1980s, Coronado, California (March 5-7, 1981) (Chicago, Ill.: American Bar Association, 1981); and Ronald Bass, "Untangling California's Environmental Maze" (Thesis, the Consortium of the California State Universities and Colleges—Environmental Planning, January 1982).

17. Environmental Law Institute, *NEPA In Action—Environmental Offices in Nineteen Federal Agencies* (Washington, D.C.: Environmental Law Institute, 1981). The Reagan Administration has drastically cut CEQ's staff and funding, thus hampering its mediating capabilities.

18. Cal. Govt. Code Ch. 4.5, sec. 65920 et seq.

19. Barbara Alexander, *The Procedural Efficiency of Maine's Environmental Permit System* (Augusta, Maine: The Maine Land and Water Resources Council, 1978), p. 105.

20. Camella Auger and Martin Zeller, *Siting Major Energy Facilities: A Process in Transition* (Boulder, Colo.: The Tosco Foundation, 1979).

21. Under Cal. Govt. Code Ch. 4.5, sec. 65943, agencies must determine within 30 days whether an application is complete. If not, the application is deemed complete. This provision was enacted in response to the Dow case as part of Assembly Bill 884 (1977).

22. Robert Bander and Abby Pirnie, *Industry Project Planning and the Permitting Process* (Washington, D.C.: U.S. Environmental Protection Agency, 1980), pp. 60-61. When passage of the EMB legislation appeared certain, the U.S. EPA responded by establishing a priority energy project (PEP) tracking system in its Washington, D.C., headquarters.

23. U.S. EPA, "Priority Energy Project Tracking System Status," Memorandum to the Administrator (August 10, 1982), p. 3; interviews with Stuart Sessions, U.S. EPA Office of Planning and Evaluation, various dates in 1982 and 1983.

24. Inside EPA (March 5, 1982), pp. 10-11; Sessions interview.

25. The grandfather provision in the final version of the Priority Energy Project Act (sec. 316) gave the EMB case-by-case authority to exempt a project from any federal, state, or local law if such law presented a "substantial impediment" to its implementation. See *Congressional Record* (June 21, 1980) (Conference Report on S. 1308), p. H5479.

26. 33 U.S.C. sec. 1316(d).

27. Bruce Beckman and Michael Prairie, "A Guide to Obtaining Required Regulatory Approvals for New Industrial Facilities in California," *San Diego Law Review*, vol. 17 (1980), pp. 1016-17.

28. Wendy U. Larsen and Charles Siemon, *Vested Rights: Balancing Public and*

Private Development Expectations (Washington, D.C.: Urban Land Institute, 1982).

29. Hopping interview.

30. 16 U.S.C. sec. 1539. For an excellent case study of how the new provisions work, see Fred Bosselman, "New Dispute Resolution Mechanisms in Federal Law," in *Real Estate Development and the Law in the 1980s* (Washington, D.C.: Urban Land Institute, 1983), p. 139. The amendments to the act are summarized in *Land Letter* (December 1982), p. 1.

31. Bosselman, "New Dispute Resolution Mechanisms," p. 143.

32. Quoted in David L. Brunner, Will Miller, and Nan Stockholm, *Corporations and the Environment: How Should Decisions Be Made* (Stanford, Calif.: Graduate School of Business, Stanford University, 1981), p. 119.

33. Ralph M. Field and Asssociates, *Perspectives on Permit Streamlining*, report prepared for the Puerto Rico Department of Natural Resources, September 1981, p. 6.

34. For a discussion of the status of state and local government environmental regulatory efforts, see *State of the Environment 1982* (Washington, D.C.: The Conservation Foundation, 1982), pp. 418-24; and *The State of the States* (Washington, D.C.: National Governors Association, 1982).

35. Interview with Dean Misczynski, director of planning, Office of Planning and Research, in Sacramento, California, August 1979.

36. For a discussion of recent cuts in the U.S. EPA budget, see *State of the Environment 1982*, pp. 389-97.

37. *The State of the States*.

38. Statement by Stanley Dempsey, vice-president of AMAX, Inc., quoted in Brunner, Miller, and Stockholm, *Corporations and the Environment*, p. 135.

39. *The State of the States*.

40. Hopping interview.

41. Richard Booth and Rosemary Nichols, "The Uniform Procedures Act: Toward a Comprehensive Permit Review System for the Department of Environmental Conservation," *Albany Law Review*, vol. 44 (April 1980).

42. Quoted in Brunner, Miller, and Stockholm, *Corporations and the Environment*, p. 164.

43. Thomas Gladwin, *Environment, Planning and the Multinational Corporation* (Greenwich, Conn.: JAI Press, 1975), p. 233.

44. Dempsey, quoted in Brunner, Miller, and Stockholm, *Corporations and the Environment*, p. 136.

45. Interviews with Howard Stafford, professor of geography, University of Cincinnati, various dates, 1982 and 1983. For a review of this point, see Andrew Huggins, *Analysis of Industry Permitting Strategies* (Washington, D.C.: U.S. EPA, Office of Policy Analysis, 1983).

46. General Accounting Office, "The Environmental Impact Statement: It Seldom Causes Long Project Delays But Could Be More Useful If Prepared Earlier," Washington, D.C. (August 9, 1977).

47. Dempsey, quoted in Brunner, Miller, and Stockholm, *Corporations and the Environment*, p. 138.

48. Robert Cahn, *Footprints on the Planet: A Search for an Environmental Ethic* (New York: Universe Books, 1978), p. 245.

200 PART III: RESPONSES TO THE NEW REALITIES

49. Ibid., pp. 246-47.
50. Three good sources, though now somewhat dated, are John Quarles, *Federal Regulation of New Industrial Plants* (Washington, D.C.: published by the author, 1979); Beckman and Prairie, "A Guide To Obtaining Required Regulatory Approvals," p. 979; and Bander and Pirnie, *Industry Project Planning*.
51. *Business Week*, October 19, 1979.
52. Discussions with Stanley Dempsey, vice-president of AMAX, Inc., in Washington, D.C., June 1979. For a detailed discussion, see Cahn, *Footprints on the Planet*, pp. 71-84.
53. Hopping interview.
54. This description and analysis is based on interviews with corporate officials, local citizens, and federal, state, and local government representatives. The site of the project was also visited by Conservation Foundation staff. For a discussion of the origins of AMAX's efforts, see Lawrence Mosher, "A Fast Track for Colorado," *National Journal* (August 2, 1980), p. 1284; John Wiebner, "A Clear Path through the Permitting Forest," *Mining Engineering* (July 1980), p. 791.
55. Oliver Wood et al., *The BASF Controversy: Employment versus Environment* (Chapel Hill, N.C.: University of North Carolina Press, 1980), p. 66.
56. Auger and Zeller, *Siting Major Energy Facilities*, p. 54.
57. Interview with Arthur Biddle, Mt. Emmons project manager, AMAX, Inc., in Denver, Colo., August 1980.
58. Address by J. Robert Ferguson, Jr., senior vice-president and assistant to the president, U.S. Steel Corporation, to a Conference Board meeting on early corporate environmental assessments, New York City (March 10, 1980).
59. The project involved was a new steel mill to be built by U.S. Steel at Conneaut, Ohio. That project has been delayed indefinitely due to slack market demand for steel. For a description of the Conneaut planning process, see Margolin, "The Two-Minute Mile," p. 17.
60. See Reginald Stuart, "Louisiana Awakening to the Need to Deal with the Pollution of Its Paradise," *New York Times*, November 27, 1982; and David Morell and Grace Singer, *Refining the Waterfront* (Cambridge, Mass.: Oelgeschlager, 1980).
61. Ibid.
62. For a good discussion of citizen participation initiatives in the siting of power plants, see David Myhra, *Public Involvement in the Introduction of Power Plants* (New York: John Wiley and Sons, forthcoming). Myhra's conclusions are summarized in *Electric World* (February 15, 1979), p. 38. He found that the failure of public participation is attributable to several management faults:
- Belief that public involvement is merely an expended version of public relations.
- Failure to comprehend the deep-seated interest the public has in the project, as well as in the environmental impact and related benefits. Project management, discovering this interest, often tries to back out, or at least limits the amount of involvement.
- Failure to organize the public-involvement program to be achieved by a certain date.
- Failure to pick a truly representative sample of the public for its public-involvement committee. Too often, this committee is made up of anti-power

people and interests, because the utility assumes that if the anti-power people can be won over by involving them in the decision-making process, then the project will likely be built as planned.
• Failure to assign the proper company talent; to identify the public's concerns; to have adequate data, reports, and surveys; and to maintain the public-involvement programs long enough to obtain a consensus.
• Failure to communicate effectively the results of the involvement program with the greater public at large.

63. Thomas Gladwin, "Patterns of Environmental Conflict over Industrial Facilities in the U.S., 1970-78," *Natural Resources Journal*, vol. 20 (April 1980), p. 254.

64. Ferguson address.

65. Statement by Timothy Shultz, county commissioner of Rio Blanco County, Colorado, at an International Bar Association seminar in Mining and Environmental Law, Washington, D.C. (April 4-8, 1982).

66. Michael Rock, "Local Government-Industry Perspectives," presented at National Conference on Coordinated Permitting Review for Energy and Mineral Resource Development, Keystone, Colorado (July 13-15, 1982).

67. For a description of the efforts in Gunnison, Colorado, to deal with growth pressures from natural resource development, see J. Ritchie Smith, "Quality-of-Life Style Study Yields Development Imperatives for Industry," *The Western Planner*, vol. 2, no. 8 (October 1981), p. 9.

68. W. Metz, "The Mitigation of Socioeconomic Impacts by Electric Utilities," *Public Utilities Fortnightly*, (September 11, 1980), reprinted in *The Western Planner*, vol. 2, no. 8 (October 1981), p. 8.

69. Gladwin, "Patterns of Environmental Conflict," pp. 259-61.

70. The following description is based on conversations with Gerald Cormick and on materials supplied by the Office of Environmental Mediation, Institute for Environmental Studies, University of Washington, Seattle, Washington. Other useful sources are "Environmental Conflict Resolution: Practioners' Perspective of an Emerging Field," *Resolve* (Winter 1981), p.1; and *Environmental Mediation: An Effective Alternative*? (Palo Alto, Calif.: Resolve, 1978).

71. Office of Environmental Mediation, "So You Are Considering Mediation . . . " (Seattle, Wash.: University of Washington, n.d.).

72. For accounts of other mediated disputes, see Allen Talbot, *Environmental Mediation, Three Case Studies: The Island, The Highway, The Ferry Terminal* (Seattle, Wash.: Institute for Environmental Studies, 1981); and Malcolm Rivkin, *Negotiated Development: A Breakthrough in Environmental Controversies* (Washington, D.C.: The Conservation Foundation, 1978).

73. Discussion with Sam Gusman, Conservation Foundation senior associate, various dates in 1983. Gusman was directly involved with this case.

74. The use of mitigation as a dispute-resolution tool is discussed in *A Handbook for States on the Use of Compensation and Incentives in the Siting of Hazardous Waste Management Facilities* (Washington, D.C.: U.S. Environmental Protection Agency, 1981); Lawrence Bacow and Debra Sanderson, *Facility Siting and Compensation: A Handbook for Communities and Developers*, MIT Energy Laboratory Working Paper No. MIT-EL-80-037WP (September 1980); Michael

O'Hare, "Not on My Block, You Don't—Facility Siting and the Strategic Importance of Compensation," *Public Policy*, vol. 28, no. 4 (1977).
75. Julia M. Wondolleck, "Bargaining for the Environment: Compensation and Negotiation in the Energy Facility Siting Process" (Master of City Planning thesis, Massachusetts Institute of Technology, 1979).
76. Bosselman, "New Dispute Resolution Mechanisms."
77. Bacow and Sanderson, *Facility Siting and Compensations*, pp. 112-19.
78. 16 U.S.C. sec. 1539.
79. U.S. EPA, "Priority Energy Project Tracking System Status.'
80. Statement by Arthur Biddle, project manager at AMAX, Inc., at National Conference on Coordinated Permitting Review for Energy and Mineral Resource Development, Keystone, Colorado (July 13-15, 1982).
81. Comment by Frank Gregg, former director of the Federal Bureau of Land Management, at a Conservation Foundation seminar on project permitting (June 1982).
82. Bander and Pirnie, *Industry Project Planning*, p. 68.
83. Ibid., p. 69.
84. For discussions of trade secret disclosure problems, see Peter Safir and Glenn Davis, "Disclosure of Pesticide Safety Data: A Viable Compromise at Last?," *Environmental Law Reporter*, vol. 12 (September 1982), p. 15017; and Rice Odell, "Agencies Curtail Access to Data and Decisions," *Conservation Foundation Letter* (November 1981). In a recent case, *Health Research Group v. FDA*, no. 82-1745 (D.C. Cir., April 15, 1983), a federal appellate court rejected the Food and Drug Administration's broad definition of trade secrets that allowed withholding of any information that would give one competitor an advantage over another. The court held that trade secrets mean something "secret [and] commercially valuable" that is related to "production."
85. *Fredericksburg Free Lance-Star*, January 17, 1982.
86. For discussion of this concept, see Safir and Davis, "Disclosure of Pesticide Safety Data"; and Odell, "Agencies Curtail Access."
87. Bander and Pirnie, *Industry Project Planning*, 65-66.
88. Yost, "Environmental Regulation—Myths, Realities and Reform," p. 258.
89. Discussions with Toby Pierce, U.S. Environmental Protection Agency, Washington, D.C., December 1981. As noted earlier, plans for the refinery have been postponed indefinitely because of falling demand for refined oil products.
90. Sherman interview.
91. S. 971, 96th Cong., 1st sess. (1979). Similar legislation was introduced in 1980 and surfaced again in 1981.
92. U.S. Bureau of Land Management, *The Nation's Public Lands* (Washington, D.C.: U.S. Bureau of Land Management, 1979), p. 43.
93. An excellent discussion of these provisions is included in A. John Davis, "Western Boom Town Financing Problems: Selected State Legislative Responses," *Journal of Contemporary Law*, vol. 5 (1979), p. 319.
94. Utah Code Ann., sec. 63-52-1 (Repl. vol. 1978).
95. Interview with Richard Moore, executive director of the Wyoming Industrial Siting Administration, in Cheyenne, Wyoming, July 1981. A thorough discussion of the Wyoming Industrial Siting Council established by the Wyoming Industrial Development Information and Siting Act can be found in Marilyn Kite, "The

Regulation of Industrial Site Selection in Western States," *Western Land Use Regulation and Mined Land Reclamation Institute* (Denver, Colo.: Rocky Mountain Mineral Law Foundation, 1979). An interesting case study of the act in action can be found in Tom Boujsty and Tom Wolf, "The Price of Prosperity," *High Country News*, vol. 14, no. 11 (May 28, 1982).

96. Williams interview.

97. For a discussion of various environmental inventories useful in project planning, see Jack McCormick, "The Effects of NEPA on Corporate Decision-making," in *Early Corporate Environmental Assessment*, p. 5.

98. The Atlantic resource surveys are described in Luther Carter, "East Coast Maps to Alert Industry to Ecology Conflict," *Science*, vol. 208 (April 11, 1980), p. 159.

99. The survey program is described in Larry Folks and Mike Scieska, "Mapping Michigan's Future," *Michigan Natural Resources Magazine*, vol. 51, no. 2 (March-April 1982), p. 12.

100. Phillip Hose, *Building an Ark—Tools for Preservation of Natural Diversity through Land Protection* (Covelo, Calif.: Island Press, 1981).

101. Interviews with Charles Forester and Gordon Jacoby, Association of Bay Area Governments, Berkeley, California, February 23, 1979.

102. For an excellent description of the Grays Harbor planning process and a critical analysis, see Nan Evans et al., *The Search For Predictability: Planning and Conflict Resolution in Grays Harbor, Washington*, no. WSG 80-5 (Seattle, Wash.: University of Washington, 1980).

103. 15 CFR Part 923, 49 Fed. Reg. 1683 (January 1975).

104. For a discussion of such cutbacks and their impacts, see *State of the Environment 1982*, p. 6.

105. Telephone interview with Richard Moore, director, Wyoming Industrial Siting Administration, August 26, 1983. According to Moore, the funds are being spent on law enforcement to guard against wildlife poachers, as well as on a literature survey and on field inventories and study.

106. For a list of other states with permit guides and manuals, see Temple, Barker, and Sloane, *Streamlining the Permitting Process*.

107. For a good survey of local reform efforts, see Charles Thurow and John Vranicar, *Procedural Reform of Local Land Use Regulation*, a paper presented at HUD National Conference on Housing Costs, Washington, D.C. (1978) (Chicago, Ill.: American Planning Association, 1978).

108. Telephone conversation with Phillip Reed, Environmental Law Institute, Washington, D.C., August 26, 1983.

109. Rotan interview. For an interesting account of how several firms are using computers to keep tabs on regulatory requirements for their facilities, see Barbara Goldsmith, "Regulatory Information Systems," *Environment, Science, and Technology*, vol. 17, no. 7 (1983), p. 311.

Chapter 7

Some Closing Thoughts

The quiet reforms, discussed in the last chapter, offer a solid alternative for rebuilding the environmental regulatory system for industrial plants into one that is more reasonable, efficient, predictable, *and* effective in protecting the environment. However, regulatory reform is no easy task under the best of circumstances. We have seen how many early efforts faltered, and today's would-be reformers, particularly at the national level, have accomplished little more than stirring up a tempest of opposition.

TIPS FOR REGULATORY REFORMERS

Despite their advantages, the quiet reforms will present substantial challenges for the regulatory community and industry. The new ways of doing things advocated in chapter 6 will be helpful in improving the process, but do not ensure success. The real world is much more complicated. There are intangible ingredients like public confidence and political support that cannot be ignored in reforming the process. The quiet reforms suggested here must therefore be implemented with great care after the proper foundation for them is laid. A few basic precautions can help keep these reforms from failing.

Maintain the System's Integrity

The success of any regulatory system depends in large part on public perception of its efficacy and support by those subject to it. If changes are to take hold, regulatory reform must not be perceived as a means for failing to enforce laws or for allowing industry to influence regulatory decisions unduly.

A modern-day example of how loss of public confidence can torpedo regulatory reform is the Reagan Administration's efforts to make the environmental regulatory system more flexible. Many knowledgeable observers agree that these reforms are, for the most part, going nowhere.[1] Why? With significant cuts in the budget of the federal Environmental Protection Agency (EPA) and a sharp scaling back of enforcement actions, efforts to make regulations more "flexible" and allow industry greater leeway in meeting standards look suspiciously like back-door attempts to amend or repeal the regulations.

Provide Necessary Checks and Balances

Although eliminating duplicative reviews of a project can speed the process, reformers must keep in mind that a streamlined system has fewer checks and balances, so any mistakes made in evaluating a project are less likely to be discovered and corrected. When the regulatory process is being redesigned to make it more efficient and effective, therefore, the need for fail-safe measures must be considered. As one report explained:

> The selection of fail-safe measures requires conscious consideration of how far the real world capability is likely to fall short of ideal . . . it is necessary to consider what each unit is likely in practice to do with the staff and funds it can reasonably expect to obtain . . . some of the measures needed to protect the integrity of the process require acknowledgment of risks that no one likes to talk about—notably the risks of erroneous decisions, misuse of decision-making power, and other irregularities. These are unpleasant to discuss and difficult to offer as justifications for steps that lengthen or delay the administrative process. But no agency can avoid these risks, and every sound regulatory process must consider them.[2]

How might a new generation of reform put in place a system that, for example, gives industry more flexibility to meet standards by its own devices rather than have the methods of control dictated by regulators? What of proposals to allow industries to certify their own compliance with regulations rather than having EPA personnel conduct inspections? Reformers must recognize the potential enforcement problems thereby created. If a novel design solution does not work, how would an agency find out about it, and what would it do to rectify the situation?

Experience shows that reforms should not rely exclusively on the professed good intentions of firms, bureaucrats, or anyone else involved in the process. As one industry environmental consultant explained: "We see that under the Reagan Administration industry is getting the signal that enforcement is not going to be as strict. Thus, they are reluctant to comply with the law if it means spending a lot of money. You simply cannot rely on them to comply on their own under such circumstances."[3]

Perhaps such a system would succeed if fail-safe measures were in place to help ensure compliance. For example, an industrial com-

SOME CLOSING THOUGHTS 207

pany might post a bond that would be forfeited if it did not comply with the law. Bonds are routinely used in analogous circumstances in the housing development industry. Local governments typically approve large housing developments, attaching conditions regarding details such as construction of roads and landscaping; then, to ensure compliance, they require the builder to post a performance bond that is refundable after the project is completed and the site inspected by a local agency. The use of bonds in an environmental regulatory context would require some changes from the land-use context— for example, the bond might be refunded in increments over the useful life of a facility—but the basic concept is not new.

Insurance might also be a helpful adjunct to "flexible" regulation. In hazardous-waste disposal, firms that run disposal sites must have insurance protecting them from liability should wastes escape and cause damage off-site. Their experience has often been that the insurance companies impose stricter operating and design standards than does the EPA simply because of the possible huge liabilities involved.[4] Similarly, some experts have proposed that the Occupational Safety and Health Administration use workers' compensation insurance as a way to reduce injury rates by granting firms with good safety records lower insurance premiums or tax deductions. Although the use of insurance has limitations (for example, society's goal may be to prevent human injury altogether, not simply compensate for it), the option is also worth exploring as an environmental regulatory tool and could help maintain public confidence in a more flexible system.

Involve All Affected Parties

In addition to checks and balances and "fail-safe" measures such as bonds and insurance, an important way to maintain confidence in the system is to involve all affected parties in the reform process— industry, public-interest groups, and agency staff. The process used by the Council on Environmental Quality (CEQ) in putting together its ambitious National Environmental Policy Act (NEPA) reform package is a textbook "how to" example. Instead of just consulting with other federal agencies or a chosen few industrialists or environmentalists, CEQ involved both critics and friends. It persuaded the U.S. Chamber of Commerce and the AFL-CIO to coordinate presenting the views of business and labor. The Natural Resources Defense Council played the same role for environmental groups. CEQ also solicited advice from a wide range of other interests, including the general public and state and local governments. Not surprisingly, five years later there is a firm consensus that the new regulations are working to make NEPA reviews more effective and expeditious.[5]

Provide Adequate Financial and Administrative Support

Closely linked to the need to maintain confidence in the system is the importance of providing adequate financial and administrative support to implement reforms. Merely switching or eliminating boxes on an organizational flowchart is a recipe for failure. There must be an assurance that the new system will produce quality work.

The Colorado Joint Review Process (CJRP) illustrates that reforms do not always come cheaply or easily. The CJRP, which now applies to only a handful of major projects, employs five full-time professional and support staff. Similarly, the Bureau of Land Management's (BLM) Office of Special Projects (OSP) found that assembling project teams that could evaluate a project quickly and effectively was fairly costly. The CJRP and OSP are two of the most ambitious of the reforms considered here, but even in lower-profile situations, reformers must make sure that already overburdened agency professionals are not simply given more work. If they are, any positive attitudes toward change will dissipate and reforms will languish, or the quality of reviews will deteriorate as regulators take on more and more work.

The EPA's effort to implement a priority energy project tracking system, described in chapter 6, is instructive. The idea of a tracking system to keep tabs on permit review for major energy projects originated with headquarters staff in Washington, D.C.[6] Although the notion seemed simple and based on common sense, its implementation was blocked by regional offices, which had the permitting information and would do most of the work. Most regional personnel viewed the proposal as an additional burden to be shouldered without additonal resources. Not until headquarters consulted with the regions in the design stage and took the lead in administering the system did the tracking initiative get off the ground. Even then, however, it was somewhat less ambitious than was originally envisioned.

In Florida, a lack of resources has hampered the state's drive to compete with Georgia's one-stop permitting system. The Florida Division of Economic Development (within the state Department of Commerce), which would like to emulate Georgia's example of helping industries identify sites with a "fighting chance" for approval, works with the state Department of Environmental Regulation to identify potential problems with proposed sites. Beyond identifying broad areas or regions, however, the system collapses, according to one official: "We just don't have the manpower or resources to narrow it down further than that. It's so complex we can't do it. It is costly, but we should be able to do the job for a few big projects."[7]

In this context, it is not surprising that Reagan Administration efforts to reform the environmental regulatory process have met with widespread skepticism. Critics ask how the federal EPA can be doing a more efficient and effective job when the agency's budget and personnel have been so drastically reduced. Similarly, while CEQ's new NEPA rules have gained widespread praise, federal agencies are finding that "lack of travel funds and the shortage of skilled personnel . . . hinder meaningful participation in scoping, particularly if a scoping meeting is involved."[8] Lack of money and trained personnel have also hampered agency efforts as cooperating agencies.

All in all, it appears that those who believe that the process can be improved and at the same time produce significant budgetary savings may not be looking at things realistically.

Maintain High Visibility

Reforms are more likely to succeed if they have another common feature: high visibility, accompanied by high-level support. This feature was crucial to the success of the CEQ reforms and other efforts. The first head of the OSP, for example, maintains that without high visibility and strong support from BLM's director, the experiment would have never got off the ground: "There was a lot of bureaucratic infighting. Many people thought the system was fine and asked why do we need a special team. The director finally changed his mind. He realized you had to show everyone that these major projects were getting special attention and high priority."[9] And in the early days of OSP, he adds, there were other crucial "status" factors.

> Access to the director was very important. You can't just bury an entity like OSP somewhere in the bureaucracy. We were also on par with the state [BLM] directors. This was important because we cooperated with state offices in reviewing projects. We also had authority and ability to go to other departments of BLM to get help.

The success of reform efforts in Colorado and California can also be partly attributed to the high level of support from the upper echelons of government. Colorado Governor Richard D. Lamm and the head of the state's Department of Natural Resources have repeatedly indicated their personal commitment to the CJRP. Likewise, in California the Office of Permit Assistance had direct access to Governor Jerry Brown.

Visibility and adequate administrative support are also keys to reforming the planning process within industry. Quiet reforms such as project teams and early environmental assessment may go for naught unless they have both the strong commitment of a firm's chief executive officer and an organizational structure that ensures the adequate evaluation of environmental issues.

A caveat is in order, however. Although high visibility and strong support from above are crucial, it is important that the reform process not be politicized. The line between visibility and playing politics can be a fine one. If a particular initiative is viewed widely as the political tool of a particular administration or the "baby" of one individual, it is more likely either to be abolished or to have its funding cut when a new administration or individual takes office. Some authorities have said that one reason the Reagan Administration killed BLM's OSP was the program's close connection with Carter Administration appointees.

Politicization will also lead some people to conclude that an effort is window dressing, not worthy of any confidence in its ability to handle complex reviews of big projects. A partial answer to this problem may be to use highly regarded career people rather than political appointees to oversee the reforms. It might also help to make high visibility of the regulatory mechanisms an everyday concern, instead of an obvious afterthought used primarily in emergencies.

Take a Comprehensive Approach

Another ingredient necessary for a reform program to reach a self-sustaining point is comprehensiveness—a broad overview of the regulatory system rather than a snapshot of problems within one agency, one level of government, or one phase of project planning. One reason many of the early one-step permitting attempts failed was their concentration on only a single, narrowly defined problem in the regulatory system: multiple state permits. These early reforms failed to consider such issues as coordination with federal and local agencies or the need for open corporate project planning.

The risk of piecemeal reform is that the changes not only may fail to improve the process but in some cases may even make things worse. As one construction industry official observed at a 1978 conference on permit reform in Hawaii:

> One by-product of our research and campaign has been to discover that there are others also actively involved in the fight . . . the major problem now will be to insure that all of these agencies work together in close coordination and cooperation to develop one common solution to the problem instead of having a proliferation of solutions which will compound the problem.[11]

Be Persistent

A final bit of advice is appropriate: even if it appears that everything has been done right and that the most impressive, progressive reform package possible has been constructed, regulatory reformers should not be dismayed if the new process fails to work perfectly the first time around.

Quiet reforms demand continuing administration, feedback from parties involved in the process, and fine-tuning to adapt to particular projects, personalities, and other variables. In other words, persistence and hard work are needed, just as in most other successful endeavors.

CONTINUING CHALLENGES

Despite negotiation, mediation, and other tools to reduce the need to wage siting battles in courtrooms, the threat of litigation will not be completely blunted in a system so dedicated as ours to judicial means of resolving disputes. There is little chance that this existing system will be completely abandoned any time soon. Similar dilemmas are inherent in almost any regulatory scheme that might conceivably gain acceptance in the United States. Some sample problem issues are:

Treating small and large businesses fairly. It is never easy to devise a regulatory process that assures adequate review of major projects while foregoing red tape that can strangle smaller firms with fewer resources. One approach that has been used in the United States and elsewhere is to subject smaller firms and their projects to less scrutiny, but the cumulative impact of such reduced stringency cannot be ignored.[12] The quiet reforms can help ease burdens on industries, large and small alike, but greater thought needs to be given to the special problems confronting smaller firms.

Resolving conflict between openness and efficiency. Government in the sunshine, with full public participation, can lead to the no-man's-land of interminable reviews, demands for more information, and a creeping paralysis in the decision-making process. Striking a proper balance will never come easy.

Reducing incentives for delay. Several informal factors may drive regulators to hold up a project that faces opposition; industry, for example, claims that no one gets kudos for approving a good project, only demerits for letting a bad one slide through. Although several of the quiet reforms such as providing better information to decisionmakers and imposing decision schedules can help reduce what industry says is institutional chickenheartedness, this problem will always exist in one form or another. Moreover, the fact that in some jurisdictions the incentives actually work the other way should not be overlooked; regulators who are not backed by a conservation-oriented constituency may find it easier to say yes than to demand more thorough review, thereby keeping the powers that be pleased.

Providing adequate funds. As financial resources shrink at all levels of government, it becomes increasingly important to give regulators adequate resources while getting the most for every regulatory dollar. Good project review and sound regulation will almost always cost more than a quick-and-dirty approach. Some say that such constraints suggest a need for concentrating resources on the biggest projects and the most serious pollution problems. The questions of who pays how much for pollution control and where money will come from for the transaction costs invite constant reexamination.

Reconcile conflicting demands for certainty and flexibility. Industry continually presses for greater certainty in the siting regulatory process but at the same time pushes for reforms to give firms greater flexibility in meeting environmental goals. However, if laws and regulations are written to give regulators more discretion in working with companies to solve particular pollution or land-use problems, that discretion can lead to uncertainty by giving regulators greater leeway in making decisions. The search for flexibility should be undertaken with the need for certainty in mind.

No reform program, no matter how smart or how ambitious, will end tensions such as these, which are built into almost any regulatory system. But their knotty quality should not be an excuse for hand wringing and inertia. There is an increasing need to experiment and

SOME CLOSING THOUGHTS 213

to test new and promising ways of controlling pollution and ensuring environmental quality.

One tool that has been little used in the environmental area is that of tax incentives or disincentives. Presidents and the Congress have used the income tax code in a wide variety of ways to encourage or discourage certain types of action or behavior. But perhaps because of their success in securing passage of strong federal "command and control" regulations, environmentalists have not widely explored the use of tax incentives.

Historic preservationists, by contrast, have used tax incentives to promote their aims, and the results have been spectacular. In the 1960s, preservationists won early legislative victories—especially, the National Historic Preservation Act of 1966—that began building a regulatory program. However, for a variety of reasons, direct federal regulation never materialized. As a result, preservationists were forced to turn to other avenues such as tax incentives, which many consider to be the linchpin of federal preservation efforts. In 1976, for example, and again in 1980, Congress enacted laws to provide tax deductions and credits to persons and businesses rehabilitating landmark properties on the National Register of Historic Places. The Department of the Interior calculates that since 1976, these new provisions have spurred landmark renovation worth over $3 billion.[13]

The possibilities of using similar tax incentives to protect environmental quality are legion. An industry that surpassed Clean Air Act standards might be given tax benefits; one that failed to meet the law might be denied certain deductions.[14] Similarly, a firm that took special steps to preserve an ecologically significant wetland on a proposed construction site might receive tax credits keyed to the amount of funds it spent on mitigative action, while one that destroyed the wetland would be penalized by losing depreciation deductions.

The country seems to be craving what Lester Thurow, professor of economics at Massachusetts Institute of Technology, calls "laetrile planning"—quick cures for complex economic problems. He concludes that unless the United States does "all those terribly difficult structural things—upgrading education, research and development, [having] a better system of strategic planning and forcing Americans to save more and consume less"—it is doomed to a cycle of quick fixes and failures.[15]

The same can be said of environmental and land-use regulatory reform. We can continue to search for wonder drugs, or we can put our noses to the grindstone and sweat and toil to bring long-lasting relief to both industry and the environment. Incremental, structural improvements offer the best hope of lasting improvements in the regulatory system. This is true not just for the federal government. State and local governments also have a valuable role to play—as laboratories for renovating regulatory approaches.

The choice should be obvious. No matter what state the economy is in, no matter whether regulatory problems might be choking a boom or stifling a recovery, the need to reform industrial plant siting and other environmental regulatory processes is pressing. The nation's citizens do not want to eliminate or gut their environmental laws and regulations. Public opinion polls—including those sponsored by industry—make that point clear.[16] But people do want the laws to work better.

That goal will not be easy. Radical change and "laetrile planning" are not the answers. Rather, "quiet" reforms—the kind suggested throughout this book—offer the most promise of protecting America's environment while also strengthening industry and the nation's economy.

REFERENCES

1. Speech by Randy Davis, Republican congressional aide, at an American Bar Association Seminar on Regulatory Reform, Washington, D.C. (October 1982); Robert Crandall, "The Environment," *Regulation: AEI Journal on Government and Society*, vol. 6, no. 1 (January/February 1982), pp. 29-32.

2. Ralph M. Field and Associates, *Perspectives on Streamlining* (Westport, Conn.: Field and Associates, 1981), pp. 18-19.

3. Telephone discussion with Barbara Goldsmith, senior program manager at Environmental Research and Technology, Inc., March 1983.

4. Discussions with Charles Humpstone, Environmental Risk Assessment Service, Boston, various dates, 1980 and 1981. The U.S. EPA originally proposed including a requirement that all disposal-site operators carry such insurance under new Resource Conservation and Recovery Act regulations, but the Reagan Administration proposed dropping such a requirement in the name of regulatory reform. In their final form, however, the new regulations do require liability insurance for owners, operators, and facilities. For a discussion of these provisions, see John Quarles, *Federal Regulation of Hazardous Wastes* (Washington, D.C.: Environmental Law Institute, 1982), pp. 104-5.

5. Nicholas Yost, "Environmental Regulation—Myths, Realities and Reform," *The Environmental Professional*, vol. 1 (1979), p. 258; and discussions with Yost and Foster Knight, former general and assistant general counsel of the Council on Environmental Quality, in Washington, D.C., various dates, 1979 and 1980.

6. Robert Bander and Abby Pirnie, *Industry Project Planning and the Permit-*

SOME CLOSING THOUGHTS 215

ting Process (Washington, D.C.: U.S. Environmental Protection Agency, 1980), p. 60.

7. Interview with Mike Goldie, Florida Division of Economic Development, Tallahassee, Florida, March 26, 1980.

8. Environmental Law Institute, *NEPA in Action—Environmental Offices in Nineteen Federal Agencies* (Washington, D.C.: Environmental Law Institute, 1981).

9. Interview with David Williams, former chief of the Office of Special Projects, Bureau of Land Management, Washington, D.C. various dates, 1981.

10. Ibid.

11. John Connell, "Perspectives on the Regulatory Maze," presented at a workshop on Government Permit Simplification, Coordination, and Streamlining sponsored by the Hawaii Department of Planning and Economic Development, Honolulu, Hawaii (July 7, 1978).

12. This approach is discussed in John Noble, John Rosenberg, and John Banta, eds., *Groping through the Maze: Foreign Experience Applied to the U.S. Problem of Coordinating Development Controls* (Washington, D.C.: The Conservation Foundation, 1976).

13. Figures provided in March 1983 by the National Park Service, Office of Cultural Resources.

14. Disincentives would have to be imposed with great care, if experience in the historic preservation field is any guide. There, tax disincentives that penalized firms for demolishing structures created a great deal of animosity toward preservation in general, even though the disincentives were quite weak.

15. Lester Thurow, "Walking on the Edge of an Economic Precipice," presented at the Annual Meeting of the American Planning Association, Dallas, Texas (May 9, 1982).

16. For a summary of recent polls, see Richard Anthony, "Trends in Public Opinion about the Environment," *Environment* (May 1982), p. 14. Another invaluable source is Robert Mitchell, *Public Opinion on Environmental Issues*, Council on Environmental Quality (Washington, D.C.: U.S. Government Printing Office, 1980).

Appendixes

Appendix A

Ranking of States' Environmental Controls

In 1983, Conservation Foundation staff constructed an index of 23 environmental and land-use indicators in an attempt to rank each state's effort to provide a quality environment for its citizens. The indicators ranged from voting records of a state's congressional delegation on selected national environmental issues to existence of state laws that address specific environmental problems. The overall focus was on regulatory programs and expenditures for environmental quality.

Each indicator was assigned a point value based on its relative importance in assessing a state's environmental efforts, as judged by Foundation staff. For example, "per capita environmental-quality-control expenditures" (Indicator 5) was assigned a maximum of 6 points, while "existence of state protection of wild, scenic, or recreation rivers" (Indicator 9) was worth a maximum of 2. The highest attainable score in this ranking was 63 points.

The rankings should be treated carefully and should not be taken as a measure of the quality of any state's environment. Moreover, although each indicator is designed to measure *effort* to protect, a state may in some instances be penalized for reasons beyond its control. For example, western states generally spend less on state parks (Indicator 12) because their residents have easy access to federal lands. And in the case of legislation, the indicators only show the existence of a law, not whether it is being implemented effectively.

However, given the total number of indicators that are used in this analysis, the ranking is a useful and accurate indication of how much effort states are making to provide a clean, healthy, and pleasing environment for their citizens.

This ranking was prepared by Christopher J. Duerksen with the assistance of William Heinemann-Ethier.

RANKING INDICATORS, POINT RANGES, AND SOURCES

Environmental Indicators

1. *Voting record of state's congressional delegation on environmental and energy issues.* Awarded: 0-4 points. Based on League of Conservation Voters ranking—0-20, 0 pts.; 21-40, 1 pt.; 41-60, 2 pts.; 61-80, 3 pts.; 81-100, 4 pts. Source: *How Congress Voted on Energy and the Environment, 1982 Directory* (Washington, D.C.: League of Conservation Voters).

2. *Existence of state environmental impact statement process.* Awarded: 0, 2, or 4 points. No statement, 0 pts.; limited statement, 2 pts.; comprehensive statement, 4 pts. Source: Nicholas A. Robinson, "SEQRA's Siblings: Precedents from Little NEPA's in Sister States," *Albany Law Review*, vol. 46 (1982), p. 1155.

3. *State planning director's ranking of priority given environmental protection by state legislature.* Awarded: 0-4 points. Source: Richard A. Mann and Mike Miles, "State Land Use Planning: The Current Status and Demographic Rationale," *American Planning Association Journal*, vol. 45 (January 1979), pp. 48, 52-53.

4. *Existence of state income tax check-off for wildlife and fisheries programs.* Awarded: 0-1 point. Sources: *National Wildlife* (August-September 1983), p. 31; *Nongame Newsletter*, publication of the Nongame Wildlife Association of North America, vol. 2, no. 4 (Summer 1983), p. 3.

5. *Per capita environmental-quality-control expenditures.* State expenditures primarily for planning, regulation and enforcement, and technical and financial assistance related to air, water, and land pollution. Awarded: 0-6 points. $0.00-$1.99, 0 pts.; $2.00-$3.99, 1 pt.; $4.00-$5.99, 2 pts.; $6.00-$7.99, 3 pts.; $8.00-$9.99, 4 pts.; $10.00-$11.99, 5 pts.; $12.00 and greater, 6 pts. Source: U.S. Department of Commerce, Bureau of the Census, *1982 State and Metropolitan Area Data Book* (Washington, D.C.: U.S. Government Printing Office).

6. *Existence of EPA-authorized state program for hazardous-waste control under the Resource Conservation and Recovery Act.* Awarded: 0-2 points. No authorization, 0 pts.; Phase I interim authorization, 1 pt.; Phase II interim authorization, 2 pts. Source: *Environmental Quality 1982: 13th Annual Report of the Council on Environmental Quality* (Washington, D.C.: U.S. Government Printing Office), p. 125.

7. *Existence of one umbrella state environmental agency with responsibility for any three of air, water, noise, and hazardous-waste pollution.* Awarded: 0-1 point. Source: *World Environmental Directory*, 4th ed. (Silver Spring, Md.: Business Publishers, Inc., 1980), pp. 395-428.

8. *Existence of tax breaks for residential and nonresidential use of solar energy.* Awarded: 0-2 points. Source: Information sheets on tax breaks for residential solar systems and nonresidential uses of solar energy, Conservation and Renewable Energy Inquiry and Referral Service (formerly the National Solar Heating and Cooling Information Center).

9. *Existence of state protection of wild, scenic, or recreation rivers.* Awarded: 0-2 points. No protection, 0 pts.; administrative protection, 1 pt.; legislative protection, 2 pts. Sources: *State Wild and Scenic River Programs: 1980* (Albany, N.Y.: New York Department of Environmental Conservation, State Rivers Program, June 1981); Bureau of Outdoor Recreation, *Report No. 43* (Spring 1977), cited in Jon A. Kusler, *Regulating Sensitive Lands* (Cambridge, Mass.: Ballinger, 1980), p. 35.

10. *Per capita expenditures for noise-control programs.* Expenditures for the development of regulations through enforcement. Awarded: 0-2 points. No expenditures, 0 pts.; more than 0.0¢-1.0¢, 1 pt.; more than 1.0¢, 2 pts. Source: *Environmental Quality 1979: 10th Annual Report of the Council on Environmental Quality* (Washington, D.C.: U.S. Government Printing Office), p. 556.

Land-Use Indicators

11. *Existence of critical-area legislation protecting wetlands or endangered-species habitat.* Awarded: 0-2 points. Source: *Environmental Quality 1979: 10th Annual Report of the Council on Environmental Quality* (Washington, D.C.: U.S. Government Printing Office), p. 490.

12. *Per capita expenditures for state parks.* Expenditures for agency operating budgets and fixed capital outlays. Awarded: 0-2 points. Less than $1.50, 0 pts.; $1.50-$4.00, 1 pt.; more than $4.00, 2 pts. Sources: *1982 Annual Information Exchange* (Santa Fe, N.M.: National Association of State Park Directors), p. 10; U.S. Department of Commerce, Bureau of the Census, *1982 State and Metropolitan Area Data Book* (Washington, D.C.: U.S. Government Printing Office), p. 448, table C, item 6.

13. *Existence of state power-plant siting law with an environmental review process.* Awarded: 0-3 points. No such law or process, 0 pts.; environmental review process, 1 pt.; power-plant siting law, 2 pts.; both law and process, 3 pts. Source: *1981 NARUC Annual Report on Utility and Carrier Regulation* (Washington, D.C.: National Association of Regulatory Utility Commissioners), p. 671.

14. *Existence of state requirements for comprehensive land-use plans and consistency of land-use decisions.* Awarded: 0-4 points. No requirements, 0 pts.; optional or limited comprehnsive planning or limited consistency requirement, 1 pt. each; mandatory comprehensive planning or nonlimited consistency requirement, 2 pts. each. Source: Edith Netter, ed., *Land Use Law: Issues for the 1980s* (Washington, D.C.: American Planning Association, February 1981), pp. 77-80.

15. *Existence of environmental protection as a stated goal in state land-use law.* Awarded: 0 or 2 points. Source: Richard A. Mann and Mike Miles, "State Land Use Planning: The Current Status and Demographic Rationale," *American Planning Association Journal*, vol. 45 (January 1979), pp. 48, 54-55.

16. *Existence of state surface-mine reclamation program approved under the Surface Mining Control and Reclamation Act (SMCRA).* Awarded: 0-3 pts. No approved state program, 0 pts.; conditionally approved program, 1 pt.; fully approved program, 2 pts.; 1 additional point awarded if state law before enactment of the SMCRA covered all minerals, regulated water flow and quality, and had requirements for conserving and replacing topsoil. Sources: U.S. Department of the Interior, *Report to the House Committee on Appropriations on the Regulatory Program of the Office of Surface Mining* (April 23, 1983); Edgar A. Imhoff, Thomas O. Friz, and James R. La Fevers, *A Guide to State Programs for the Reclamation of Surface Mined Areas*, U.S. Geological Survey Circular No. 731 (Washington, D.C.: U.S. Government Printing Office, 1976).

17. *Existence of state floodplain law regulating development in floodways and floodplains.* Awarded: 0-2 points. No regulation, 0 pts.; regulation of floodways, 1 pt.; regulation of floodplains, 2 pts. Sources: U.S. Army, Corps of Engineers, *Perspective on Flood Plain Regulations for Flood Plain Management* (Washington, D.C.: U.S. Government Printing Office, June 1976), pp. 90-101; *Environmental Quality 1979: 10th Annual Report of the Council on Environmental Quality* (Washington, D.C.: U.S. Government Printing Office), p. 490.

18. *Existence of specific state land-use policies or laws.* Fourteen land-use variables were analyzed to determine the extent of state involvement in land-use decision-making. Award: 0-6 points. Based on Dillard composite score—0-1 policies or laws, 0 pts.; 2-3, 1 pt.; 4-5, 2 pts.; 6-7, 3 pts.; 8-9, 4 pts.; 10-11, 5 pts; 12-14, 6 pts. Source: Jan E. Dillard, "State Land-Use Policies and Rural America," in Robert Browne and Dan Hadwiger, eds., *Rural Policy Problems: Changing Dimensions* (Lexington, Mass.: Lexington, 1982).

19. *Adoption of an aesthetic rationale, standing alone, to support use of the police power.* Awarded: 0-2 points. Rejection of aesthetics-only rationale or no reported court cases, 0 pts.; adoption of rationale based on court dictum, 1 pt.; adoption of rationale based on court holding, 2 pts. Sources: Samuel Bufford, "Beyond the Eye of the Beholder: A New Majority of Jurisdictions Authorize Aesthetic Regulation," *University of Missouri–Kansas City Law Review,* vol. 48 (1980), p. 125; Clan Crawford, Jr., "The Metromedia Impact," *Zoning and Planning Law Reports,* vol. 6, no. 2 (February 1983).

20. *Per capita expenditures for natural resources, parks and recreation, sewerage, sanitation, and housing and urban renewal.* "Natural resources" include conservation and development of agriculture, fish and game, forestry, and other soil and water resources; "sanitation" includes street cleaning, collection and disposal of garbage and other waste, and sanitary and storm sewage disposal facilities and services; "housing" covers housing and redevelopment projects and any promotion or support of private housing and redevelopment activities. Awarded: 0-2 points. Less than $30.00, 0 pts.; $30.00-$59.99, 1 pt.; $60.00 and greater, 2 pts. Source: U.S. Department of Commerce, Bureau of the Census, *1982 State and Metropolitan Area Data Book* (Washington, D.C.: U.S. Government Printing Office), pp. 513, table C, item 1113, and 448, item 6.

21. *Existence of EPA-approved state solid (nonhazardous) waste management plan under the Resource Conservation and Recovery Act.* Awarded: 0-2 points. No plan or in draft stage, 0 pts.; plans adopted by state and submitted to EPA for approval, 1 pt.; plan approved or partially approved by EPA, 2 pts. Source: *Environmental Quality 1982: 13th Annual Report of the Council on Environmental Quality* (Washington, D.C.: U.S. Government Printing Office), p. 121.

22. *Existence of agricultural preservation tools.* Includes agricultural districting, enabling legislation for agricultural zoning,

and purchase or transfer of development rights. Awarded: 0-3 points. Source: *Farmland*, newsletter of the American Farmland Trust, vol. 3, no. 2 (1983), p. 4.

23. *Existence of a state register of historic places or recognition of conservation restrictions.* A conservation restriction is a less-than-fee property interest, including a restriction, easement, covenant, or condition. Awarded: 0-2 points. Source: James P. Beckwith, Jr., "Preservation Law 1976-80: Faction, Property Rights, and Ideology," *North Carolina Central Law Journal*, vol. 11 (1980), p. 276; Comment, "Conservation Restrictions: A Survey," *Connecticut Law Review*, vol. 8 (1975-76), p. 383.

State-by-State Breakdown by Indicator

State	1 congressional voting record (0-4)	2 state EIS process (0-4)	3 priority envir. protection (0-4)	4 wildlife tax check-off (0-1)	5 $ for envir. quality control (0-6)	6 hazardous waste program (0-2)	7 umbrella envir. agency (0-1)	8 solar energy tax breaks (0-2)	9 protection of rivers (0-2)	10 $ for noise pollution (0-2)	11 endangered species, wetlands (0-2)	12 $ for state parks (0-2)	13 power plant siting law (0-3)	14 comprehensive land-use planning (0-4)	15 envir. is stated goal (0-2)	16 surface mine reclamation (0-3)	17 flood plain protection (0-2)	18 Dillard study (0-6)	19 aesthetics and zoning (0-2)	20 $ for natural resources (0-2)	21 solid waste program (0-2)	22 agricultural preservation (0-3)	23 historic preservation (0-2)	TOTAL (0-63)
Alabama	1	0	0	1	0	1	0	0	1	0	0	1	0	0	0	1	0	1	0	0	2	0	0	10
Alaska	0	0	0	1	6	0	1	2	0	1	1	2	0	2	2	2	0	2	0	1	0	0	0	23
Arizona	1	2	1	0	1	1	1	2	0	2	0	2	3	0	0	0	2	1	0	0	2	1	1	24
Arkansas	1	0	1	1	1	2	1	0	0	0	0	2	2	4	0	3	2	2	1	0	2	0	2	27
California	2	4	4	0	3	1	0	1	2	1	2	2	3	4	2	0	1	4	2	2	2	2	2	46
Colorado	2	0	2	1	2	0	1	1	0	1	1	1	0	0	0	1	1	5	2	1	1	1	1	26
Connecticut	3	4	2	0	5	2	1	0	1	0	2	0	3	0	0	0	0	4	1	0	2	0	2	32
Delaware	3	2	3	1	0	1	1	2	2	1	2	1	0	1	0	0	0	5	2	1	0	0	0	29
Florida	2	0	3	0	1	1	1	0	0	0	2	2	1	4	0	1	0	2	2	0	2	2	1	31
Georgia	2	2	3	0	2	2	1	2	1	2	2	1	2	0	0	0	2	3	0	0	2	1	0	25
Hawaii	3	4	3	1	4	0	1	0	0	1	2	0	1	3	0	1	2	3	1	2	0	1	1	34
Idaho	0	0	1	1	1	1	1	2	1	0	0	2	0	3	0	0	0	3	0	0	1	1	0	16
Illinois	2	0	3	1	1	1	1	0	0	2	1	0	1	3	1	2	2	3	1	0	2	0	1	28
Indiana	3	4	2	0	1	0	0	2	2	1	1	1	3	0	0	1	2	1	1	2	1	3	1	36
Iowa	3	0	3	1	0	1	1	0	0	0	0	1	2	3	0	2	0	2	1	0	0	1	1	29
Kansas	1	0	1	1	4	0	0	2	2	0	0	0	0	0	0	1	0	4	1	0	2	1	0	23
Kentucky	2	2	2	1	4	1	1	2	2	2	2	2	3	1	2	2	2	3	1	1	2	0	2	34
Louisiana	2	0	1	1	5	1	0	0	2	0	1	1	2	2	0	1	0	6	1	2	2	1	1	21
Maine	3	4	3	0	0	0	1	2	2	0	0	1	3	0	0	2	2	5	0	0	2	0	1	32
Maryland	3	0	1	1	4	1	0	2	2	2	2	1	0	3	2	1	1	2	2	1	2	1	0	37
Massachusetts	4	4	2	1	5	1	1	0	2	2	1	1	3	0	0	2	2	5	1	0	2	1	2	44
Michigan	3	4	4	0	2	0	0	1	2	0	2	0	0	3	2	0	2	2	0	0	2	0	1	30
Minnesota	2	4	4	1	6	2	0	0	2	0	2	2	3	0	0	2	0	5	1	1	2	2	1	47
Mississippi	1	2	0	0	1	0	1	0	0	0	0	2	0	3	0	2	0	2	0	0	2	0	0	15
Missouri	2	0	1	0	0	1	0	0	0	0	1	1	0	0	0	2	0	2	1	0	1	1	1	14

State																					Total
Montana	2	4	2	1	3	1	1	2	1	3	1	0	2	0	3	2	2	1	0		37
Nebraska	1	2	1	0	1	1	1	2	0	1	1	2	0	1	2	0	1	1	0		22
Nevada	1	3	0	1	0	0	0	0	2	3	2	0	0	0	3	0	1	0	1		22
New Hampshire	3	0	0	4	1	0	1	1	2	2	0	0	0	2	2	1	1	1	1		21
New Jersey	3	4	2	1	6	0	1	2	2	1	2	1	0	2	5	2	1	0	3	2	45
New Mexico	0	0	0	1	0	1	0	1	0	1	2	1	0	2	2	0	1	0	0		18
New York	3	4	1	1	3	0	1	2	0	3	0	2	1	0	2	2	0	0	2	2	37
North Carolina	1	4	1	1	2	2	2	1	0	1	0	0	0	2	2	0	1	2	0	1	25
North Dakota	3	0	1	0	1	1	1	0	0	3	1	0	0	2	2	1	1	1	1		22
Ohio	2	0	2	1	6	0	0	1	0	3	0	2	0	0	2	2	1	2	1	1	30
Oklahoma	2	0	2	1	0	2	0	2	2	0	0	2	2	0	1	0	2	2	0	0	19
Oregon	3	0	2	1	4	1	2	2	2	3	3	0	0	2	5	2	1	2	1	3	42
Pennsylvania	2	0	2	1	2	0	1	2	2	0	0	2	1	1	3	1	0	0	1	1	28
Rhode Island	4	2	2	0	4	1	2	0	0	0	1	0	0	0	4	0	0	1	1		26
South Carolina	1	0	4	1	1	2	1	2	2	0	0	0	2	0	3	0	0	1	0		25
South Dakota	1	4	3	0	1	0	1	2	1	3	2	0	1	0	5	3	1	0	1	1	30
Tennessee	2	0	2	0	1	1	2	0	2	0	0	2	1	0	3	0	0	2	0	1	23
Texas	1	2	2	0	1	1	2	1	1	1	0	1	0	1	4	0	1	0	0	1	22
Utah	0	2	1	1	0	0	2	0	0	0	0	2	0	1	5	2	0	1	1	1	23
Vermont	3	0	2	0	1	0	2	0	0	3	3	0	3	0	5	0	1	0	1	1	32
Virginia	0	4	1	1	0	0	2	0	2	3	0	2	0	2	4	0	1	0	0	1	28
Washington	3	4	0	2	1	1	2	2	2	3	0	0	2	1	5	1	1	1	1	3	39
West Virginia	1	0	4	1	0	0	0	0	2	0	1	2	2	1	2	1	1	1	0	1	23
Wisconsin	3	4	2	2	1	1	1	2	0	3	0	2	2	2	3	2	0	0	2	1	37
Wyoming	2	0	4	0	2	0	0	0	2	2	1	2	0	0	2	0	2	0	0	1	23

How Individual States Fared

State	Score	Rank
Alabama	10	50
Alaska	23	33
Arizona	24	32
Arkansas	27	26
California	46	2
Colorado	26	27
Connecticut	32	14
Delaware	29	21
Florida	31	17
Georgia	25	29
Hawaii	34	12
Idaho	16	47
Illinois	28	23
Indiana	30	11
Iowa	29	21
Kansas	23	33
Kentucky	34	12
Louisiana	21	43
Maine	32	14
Maryland	37	7
Massachusetts	44	4
Michigan	30	18
Minnesota	47	1
Mississippi	15	48
Missouri	14	49
Montana	37	7
Nebraska	22	39
Nevada	22	39
New Hampshire	21	43
New Jersey	45	3
New Mexico	18	46
New York	37	7
North Carolina	25	29
North Dakota	22	39
Ohio	30	18
Oklahoma	19	45
Oregon	42	5
Pennsylvania	28	23
Rhode Island	26	27
South Carolina	25	29
South Dakota	30	18
Tennessee	23	33
Texas	22	39
Utah	23	33
Vermont	32	14
Virginia	28	23
Washington	39	6
West Virginia	23	33
Wisconsin	37	7
Wyoming	23	33

Ranking of States' Environmental Controls

State	Rank	Score
Minnesota	1	47
California	2	46
New Jersey	3	45
Massachusetts	4	44
Oregon	5	42
Washington	6	39
Maryland	7	37
Montana	7	37
New York	7	37
Wisconsin	7	37
Indiana	11	36
Hawaii	12	34
Kentucky	12	34
Connecticut	14	32
Maine	14	32
Vermont	14	32
Florida	17	31
Michigan	18	30
Ohio	18	30
South Dakota	18	30
Delaware	21	29
Iowa	21	29
Illinois	23	28
Pennsylvania	23	28
Virginia	23	28
Arkansas	26	27
Colorado	27	26
Rhode Island	27	26
Georgia	29	25
North Carolina	29	25
South Carolina	29	25
Arizona	32	24
Alaska	33	23
Kansas	33	23
Tennessee	33	23
Utah	33	23
West Virginia	33	23
Wyoming	33	23
Nebraska	39	22
Nevada	39	22
North Dakota	39	22
Texas	39	22
Louisiana	43	21
New Hampshire	43	21
Oklahoma	45	19
New Mexico	46	18
Idaho	47	16
Mississippi	48	15
Missouri	49	14
Alabama	50	10

STATES' ENVIRO

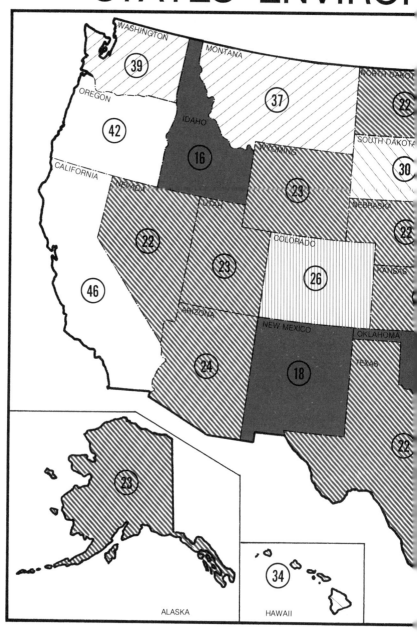

NTAL CONTROLS

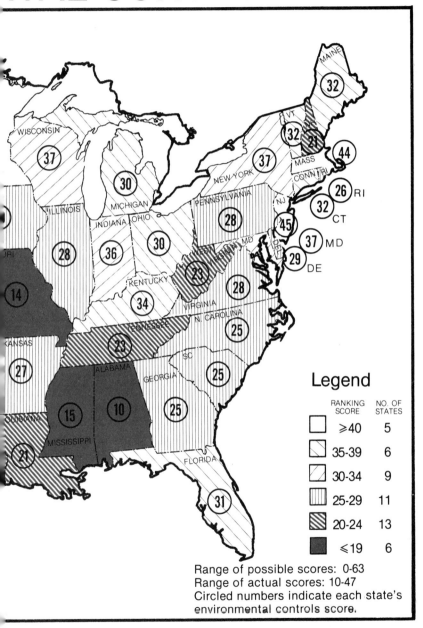

Legend

RANKING SCORE	NO. OF STATES
≥40	5
35-39	6
30-34	9
25-29	11
20-24	13
≤19	6

Range of possible scores: 0-63
Range of actual scores: 10-47
Circled numbers indicate each state's environmental controls score.

Appendix B

The Remodeled Colorado Joint Review Process

STAGE I—DECISION

Since the JRP [Joint Review Process] is a voluntary process, Stage I may only be initiated by the proponent of a project. Once a request has been made by a project proponent, a decision must be made by the executive directors of relevant state agencies based on consultation with other relevant parties (e.g., agencies and individuals) about whether the project qualifies for a Joint Review. Stage I procedures are initiated by the proponent. Three decision-making criteria influence screening decisions.

1. A proposed project should fit within the definition of a "Major Energy and Mineral Resource Development Project" as described in Stage I, Step 2.
2. A proposed project should be offered for Joint Review in an early project phase. Project phases include designation, exploration, design/feasibility, construction, operation, and post-operation.
3. State agencies should have the staff capability to meet commitments inherent in a JRP. Since these agencies would also have to review major projects under conventional review processes, some attention should be given to the time and staff workload savings using an efficient coordinated review process for a particular subject.

It is anticipated that Stage I should be completed in 31 to 38 days.

STAGE II—ORGANIZATION

Once a decision has been made to conduct the JRP, several organizational activities must occur to provide a sound framework for Stage III (implementation of a decision schedule). Stage II activities include:

1. Designation of a lead state agency by the Governor;
2. Negotiations with the other involved levels of government to obtain designations of lead agencies;

3. Signing a Joint Statement which publicly commits the federal, state, and local levels of government to participate fully in the JRP;
4. Conducting several JRP Team meetings (about six are recommended) for the purpose of organizing and planning Stage III events;
5. Signing a Statement of Responsibilities which details the responsibilities of each agency and the proponent;
6. Conducting several major public participation events (four are recommended); and
7. Preparation of the JRP Decision Schedule.

It is anticipated that Stage II can be completed in about eight months.

STAGE III—IMPLEMENTATION

The Decision Schedule prepared in Stage II will provide detailed guidelines for coordinating regulatory processes, public participation events, and JRP administrative processes into one logical, interrelated sequence of events. Generally, each model will be comprised as follows:

1. Project phases will serve as the organizational framework within which the scheduling of specific development, regulatory, public participation, and JRP activities will be arranged.
2. The proponent's schedule of specific project planning, design, and feasibility activities (i.e., preparation and completion of various reports and studies, anticipated dates of construction commencement, and anticipated dates of operation commencement) will serve as the foundation for the coordinated decision schedule. These activities will be correlated with the government decisionmaking processes and JRP activities and scheduled in terms of linear time. This proposed schedule will provide a more specific basis for scheduling various governmental actions, public participation events and JRP activities.
3. Governmental actions will be correlated, to the extent possible, in accordance with the proponent's proposed schedule, prerequisite governmental actions, and required time frames. Specific components of governmental decisionmaking processes will be scheduled to enable coordination of public notices, public hearings, submission requirements (e.g., various environmental reports), and final decisions. The goal of such coordination

is to attempt to sequence events such that final government decisions occur at or near the same time, thus enabling the proponent to make timely corporate decisions about the proposed project.
4. Public participation events will be incorporated in this master schedule at critical points. Some public participation activities are already required by law or regulation, others will be suggested for consideration during the JRP. Suggested JRP public participation events are not intended to be scheduled so as to conflict with legally required events, but rather to enhance and amplify public input during the review process.
5. JRP meetings will continue on a regular schedule in order to ensure optimum coordination of governmental, public, and corporate activities; to ensure that delays in governmental decision-making are minimized; and to provide stability to the coordination effort.

This material is excerpted from *Colorado's Joint Review Process for Major Energy and Mineral Resource Development Projects* (Denver, Colo.: Colorado Department of Natural Resources, 1980). Work on this manual was supported by the U.S. Department of Energy, Office of the Environment, under contract no. 03-80EV10233.

Index

A

Administrative Procedures
  Act, 39, 126
AFL-CIO, 207
Alabama
  environmental controls
    rank, *224, 226, 227*
Alaska, 26, 166, 189
  environmental controls
    rank, *224, 226, 227*
Alpetco refinery, 189
AMAX, Inc., 84, 171-172,
    173, 175, 177, 179
  rating of CJRP by, 135-140
  testing of CJRP by,
    131-135
American Bar Association,
    116, 122-123
American Petroleum Institute,
    116
Arizona
  environmental controls
    rank, *224, 226, 227*
Arkansas, 68
  environmental controls
    rank, *224, 226, 227*
Association of Bay Area
  Governments, 194

B

Barlöcher CWM, 101, 121
  case study of chemical
    plant siting by, 89-95
BASF, 59, 176
  case study of plant siting
    by, 4-7
Bay Area Air Pollution
  Control District, 25-26

Bureau of Land Management,
  32, 117, 151-152, 153, 161,
  191, 193
  see also Office of Special
    Projects
Bonds, 206-207
British Petroleum, 27
Brown, Jerry, 19-20, 21-22,
  210
Bumpers amendment, 127-130
Bumpers, Dale, 127-130
Business Roundtable Study,
  49, 82, 85, 86-87n
Business Week, 7

C

California, 50, 62, 68, 154,
  155, 157, 159, 174, 180,
  189, 210
  and Dow Chemical
    Company Solano plant
    siting, 18-26
  environmental
    controls rank, *224, 226,
    227*
  government site selection
    in, 125
  and Pactex project siting,
    26-35
California Environmental
  Quality Act, 24
California Office of Permit
  Assistance, 22, 157, 158,
  189
California Office of Planning
  and Research, 24, 32
California Supreme Court,
  25-26

Carter, Jimmy, 42, 112, 117, 128, 134, 210
Chevron, 139-140
Citizens Against the Refinery's Effects, 36, 39
Citizens Task Force on SOHIO, 28, 29, 31
Civiletti, Benjamin, 128
CJRP, 152, 189, 208
  as alternative to energy mobilization board, 138
  conception of, 130-131
  lead agencies in, 133
  project decision schedules in, 160-161
  rating of, by participants, 135-140
  remodeled process of, 230-232
  use of, by AMAX, Inc., 131-140
Clean Air Act, 12, 14, 22, 25, 42, 49, 57, 58, 68-70n, 80, 81, 82, 162, 192, 213
Clean Water Act, 12, 39, 162, 163
Coal Creek, 81
Coastal Zone Management Act, 117
Colorado, 84, 112, 117, 122, 152, 155, 166, 175, 179, 180, 189, 190, 192, 196, 210
  environmental controls rank, *224, 226, 227*
  joint review process of, 130-140
Colorado Department of Natural Resources, 112-115, 131-140, 161
Colorado Department of Public Health, 135
Colorado Joint Review Process. *See* CJRP
Colstrip, 81, 83
Compensation. *See* Mitigation

Competition, foreign, 50-56, 88-102
"Conditional" permits, 186-187
Congressional Research Service, 39
Connecticut
  environmental controls rank, *224, 226, 227*
Consolidated Edison, 182-183
Consolidation, 109-115
Construction Industry Legislative Organization of Hawaii, 80
Copper Environmental Equalization Act, 50
Cormick, Gerald, 182
Cost competition
  factor in regulatory misconceptions, 50-70
  see also Economics
Council on Environmental Quality, 40, 42, 49, 85, 86, 151, 157-158, 189, 196, 207-208, 209
Cox Enterprises, 35, 42

D

Delaware
  environmental controls rank, *224, 226, 227*
Dispute settlement techniques
  innovations in, 180-184
  mediation, 182-183
  mitigation, 183-184
Dow Chemical Company, 35, 43, 57, 87, 88, 89, 95, 101, 121, 123, 125, 151, 153, 155, 157, 159, 166, 169, 171, 175, 178, 180, 181, 186, 189, 190, 192
  case study of Solano plant siting, 18-26
  outcome of Solano plant siting, 21-22

regulatory difficulties of, in
 Solano plant siting, 22-26
Dupont Chemical Company,
 50, 89

E

Eagleton, Thomas, 129
Economics
 and environmental
  protection costs, 49-50
 financing of regulatory
  reform, 208-209
 foreign and domestic cost-
  competition factors, 52-56,
  58-70
 of Reagan Administration,
  70n
 and regulatory personnel
  problems, 166-168
 tax incentives for
  compliance, 213-214
El Paso Natural Gas
 Company, 30, 34
Employment shifts. *See*
 Pollution havens
Energy mobilization board,
 117, 122, 134, 159-160, 163,
 180
Endangered Species Act,
 164-165, 184
Environmental assessment,
 early, 169-172
Environmental controls,
 ranking of states by,
 218-223, *224-227*
Environmental impact
 statement. *See* National
 Environmental Policy Act
Environmental reconnaissance
 statements, 172-174
Environmental regulation
 process. *See* Permitting
 process
EPA. *See* U.S. Environmental
 Protection Agency
Escort services, 156-157

Europe
 case studies of industrial
  siting in, 89-100
 government site selection
  in, 125
 myth of better permitting
  process in, 88-89
Evans, John, 35
Executive Order 12044,
 126-127

F

Federal Energy
 Administration, 29, 32-33
Federal Power Commission,
 86
Florida, 113, 154, 196, 208
 environmental controls
  rank, *224, 226, 227*
 super-siting agency in,
  118-119
Florida Department of
 Environmental Regulation,
 118-119
Florida Industrial Siting Act
 (1979), 119, 164, 168
Florida Power Plant Siting
 Act, 118-119, 154-155
Fortune 500, 59
Freedom of Information Act,
 187
Frostbelt-Sunbelt shift. *See*
 Pollution havens, myth of
 domestic

G

General Accounting Office,
 170
Georgia, 5, 61, 152, 156-157,
 208-209
 agency consolidation
  program of, 112-115
 environmental controls
  rank, *224, 226, 227*
Georgia Conservancy, 114

Georgia Department of
  Industry and Trade, 114,
  156-157
Georgia Environmental
  Protection Division,
  112-115, 152, 156-157
Godwin, Mills, 38
Goodyear Tire and Rubber
  Company, 2
Government perspective. *See*
  Permitting process
Grandfather clauses, 162-165
Grayrocks Dam, 183
Green Party, 99-100

H

Hampton Roads Energy
  Company. *See* HRECO
Hart, Gary, 190
Hawaii
  environmental controls
  rank, *224, 226, 227*
Hearing, joint or
  consolidated, 154-156
Hickel, Walter, 6
HRECO, 43, 59, 81, 123,
  158, 170, 192, 193
  case study of Portsmouth
    refinery by, 35-43
  environmental objections
    to, Portsmouth refinery
    project, 37
  outcome of, Portsmouth
    refinery siting, 38-39
  regulatory difficulties with
    Portsmouth refinery, 39-43

I

Idaho
  environmental controls
  rank, *224, 226, 227*
Illinois, 3
  environmental controls
  rank, *224, 226, 227*

Illinois Institute of Natural
  Resources, 70
Indiana
  environmental controls
  rank, *224, 226, 227*
Industrial Development
  Research Council, 122
Industrial siting process
  in 19th century, 1-2
  in the 1960s, 2-7
  changes in, for the 1980s,
    43-44
  classic location theory of, 2
  cost-competition factor in,
    49-71
  industry perspective of,
    8-11
  parallels between U.S. and
    foreign, 100-102
  problems associated with,
    149
  reform of, in the 1970s,
    109-142
  regulatory perspective of,
    11-15
  see also Permitting process;
    Project planning
Industry perspective. *See*
  Project planning
Industry, pollution-intensive
  environmental regulation
    index for, 63, *64-67*
  investment and trade
    patterns of, 54-56
Inflation, 49-50n
Insurance, 207
Intermountain power project,
  83
Internal Revenue Service, 188
Investments, by foreign
  countries, *69*
Iowa
  environmental controls
  rank, *224, 226, 227*

J

Joint review process, 130-140

K

Kaiparowits, 81
Kansas
　environmental controls rank, *224, 226, 227*
Kelly Springfield Tire Company
　case study of, 2-4
Kentucky
　environmental controls rank, *224, 226, 227*
Kimberly-Clark, 113

L

Lamm, Richard, 117, 130, 131, 210
Laxalt, Paul, 126, 127-130
Lead agencies, 151-152
　in CJRP, 133
League of Women Voters, 28, 33
Legislation. *See* Regulations
Lever, William, 1, 2
Location theory, 2
　erosion of classic, 57
　relationship to environmental laws, 59-70
Louisiana, 83, 177-178
　environmental controls rank, *224, 226, 227*
Louisiana Shellfish Producers Association, 83

M

Maine, 68, 111, 159
　environmental controls rank, *224, 226, 227*
Maryland, 3, 154
　environmental controls rank, *224, 226, 227*
　government site selection in, 124
Maryland Department of Natural Resources, 124
Maryland Power Plant Siting Act, 124
Massachusetts, 61
　environmental controls rank, *224, 226, 227*
Mead Corporation, 173
Mediating bodies, interagency, 157-158
Mediation, 182-183
Michigan, 193-194, 195
　environmental controls rank, *224, 226, 227*
Michigan Department of Natural Resources, 194
Miller Brewing Company, 113
Mineral Lands Leasing Act, 191
Minnesota, 136
　environmental controls rank, *224, 226, 227*
Minnesota Siting Act, 121
Mississippi
　environmental controls rank, *224, 226, 227*
Missouri
　environmental controls rank, *224, 226, 227*
Mitigation, 183-184
Montana, 123
　environmental controls rank, *225, 226, 227*
Montana Major Facility Siting Act, 123

N

National Coal Association, 86
National Council of the Paper Industry for Air and Stream Improvement, 81, 84

National Electric Reliability
Council, 86
National Environmental
Policy Act, 7, 12, 34, 39,
40, 151, 158, 177, 207, 208
National Historic Preservation
Act (Section 106 process),
116, 213
National Marine Fisheries
Service, 36, 83
National Oceanic and
Atmospheric
Administration, 37, 41,
117, 193
National Wildlife Federation,
39
Natural Resources Defense
Council, 207
Nature Conservancy, 194
Nebraska, 183
environmental controls
rank, *225, 226, 227*
NEPA. *See* National
Environmental Policy Act
Nevada, 126
environmental controls
rank, *225, 226, 227*
New Hampshire, 154
environmental controls
rank, *225, 226, 227*
New Jersey
environmental controls
rank, *225, 226, 227*
New Mexico, 61
environmental controls
rank, *225, 226, 227*
New York, 123, 183
environmental controls
rank, *225, 226, 227*
New York Uniform
Procedures Act, 168
New York University
Graduate School of
Business Administration, 51
North Carolina, 3-4, 5, 68
environmental controls
rank, *225, 226, 227*

North Dakota, 121
environmental controls
rank, *225, 226, 227*

O

Office of Environmental
Mediation, 182
Office of Management and
Budget, 115-116, 126
Office of Special Projects,
117, 151-152, 153, 161, 191,
208, 209-210
Offsets, 22, 68-70n, 89
Ohio, 84, 153, 179
environmental controls
rank, *225, 226, 227*
Oklahoma
environmental controls
rank, *225, 226, 227*
One-stop permitting process,
109-115
Oregon, 68, 154, 182, 187
environmental controls
rank, *225, 226, 227*
Outline approval, 186

P

Pactex project
case study of, of SOHIO,
26-35
environmental concerns
about, 28
outcome of, 29-31
regulatory difficulties of,
31-33
significance of regulatory
delays of, 33-35
see also SOHIO
Pennsylvania, 153, 179
environmental controls
rank, *225, 226, 227*
PEP, 161-162, 208
Permit registers, 196

Permitting process
  checks and balances in, 206-207
  of CJRP, 131-140, 230-232
  consolidated, in Georgia, 112-115
  consolidation, 109-115
  delays of, in Pactex project, 33-35
  in Dow Solano plant site case, 20
  effect of, on project planning, 82-88
  expert project teams in, 152-154
  government site selection in, 123-125
  grandfather clauses in, 162-165
  growth of, 79-81
  for HRECO Portsmouth refinery, 36-37
  interagency mediating bodies in, 157-158
  joint or consolidated hearings in, 154-156
  lead agencies in, 151-152
  and local governments, 190-192
  management techniques for, 150-168
  myth of better, in foreign countries, 88-89
  myth of strangling industrial development by, 81-82
  need for predictability in, 158-165
  one-stop, 109-115
  of Pactex project, 27-28
  perception of complexity in, 79-103
  for power plants in Florida, 118-119
  preapplication permit identification in, 12-13
  preparation of permit applications, 13
  problems with European, 121-122
  project managers and escort services in, 156-157
  private ruling in, 188
  project decision schedules in, 159-161
  project tracking systems in, 161-162
  reading room option for, 188
  regulators' role in, 11-15
  regulatory personnel problems in, 165-168
  restraining regulators in, 126-130
  scoping in, 189
  site surveys in, 193-196
  submission and processing of applications, 13-15
  truncation of, 115-123
  ways to mesh with project planning, 185-187
Pollution havens
  myth of domestic, 56-71
  myth of foreign, 50-56
Portsmouth refinery. See HRECO
Preapplication meetings, 186
President's Council of Economic Advisors, 43
Prevention of significant deterioration regulations. See PSD
Priority energy project system. See PEP
Project decision schedules. See Schedules, project decision
Project managers, 156-157
Project planning
  determination of need/appropriation of capital, 8-10

dispute settlement in,
  180-184
by Dow for Solano plant
  site, 18-19
early environmental
  assessment in, 169-172
early public involvement in,
  176-178
effect of permitting process
  on, 82-88
environmental reconnaissance
  statements in, 172-174
of HRECO Portsmouth
  refinery, 35-36
and local governments,
  190-192
organization and
  management of, 174-176
permit registers for, 196
and permitting process
  delays, 80-82
plant design, construction,
  and operation, 11
plant site search, 10-11
preapplication permit
  identification, 12-13
private ruling in, 188
problems with, in
  Portsmouth refinery, 39-40
quality-of-life impacts in,
  178-180
reading room option for,
  188
scoping in, 189
shortcomings of, in Pactex
  project, 34-35
site surveys in, 193-196
steps in, 9
team for, 174-176
trade secrets in, 187-189
ways to mesh with
  permitting process, 185-187
see also Permitting process
Project teams, 152-154,
  174-176
Project tracking systems. See
  Tracking systems, project

PSD, 68-70n, 83-84, 86-87n,
  89
Public involvement
  in project planning, 176-178
  in regulatory reform,
    207-208
Puerto Rico Environmental
  Quality Board, 166

R

Reagan Administration, 152,
  163, 166-167, 210
  budget cuts by, 70n
  citizen involvement in
    regulatory policy of,
    115-116
  and pollution-control
    oversight, 57
  public confidence in, 205,
    206, 209
  restraining of regulators by,
    126-130
  use of truncation approach
    by, 117
Red tape. See Permitting
  process
Regulations
  Administrative Procedures
    Act, 39, 126
  Bumpers amendment,
    127-130
  California Environmental
    Quality Act, 24
  Clean Air Act, 12, 14, 22,
    25, 42, 49, 57, 58, 68-70n,
    80, 81, 82, 162, 192, 213
  Clean Water Act, 12, 39,
    162, 163
  Coastal Zone Management
    Act, 117
  Copper Environmental
    Equalization Act, 50
  Endangered Species Act,
    164-165, 184
  Executive Order 12044,
    126-127

Florida Industrial Siting
  Act, 164-168
Florida Power Plant Siting
  Act, 118-119, 154-155
interpretation of, 58
Maryland Power Plant
  Siting Act, 124
National Environmental
  Policy Act, 7, 12, 34, 39,
  40, 158, 177, 207, 208
New York Uniform
  Procedures Act, 168
offsets, 22, 68-70n, 89
PSD, 62, 68-70n, 83-84,
  86-87n, 89
Regulatory Procedures Act,
  129
Regulatory Reform Act,
  126-130
Resource Conservation and
  Recovery Act (RCRA), 162
Rivers and Harbors Act, 39
Surface Mining Control
  and Reclamation Act, 192
Toxic Substances Control
  Act, 162
Washington Environmental
  Coordination Procedures
  Act, 155
Water Pollution Control
  Act, 50
Water Quality Act (1965), 6
Williamson Act
  (California), 20
Regulators
  adequate information for,
    185-192
  fee charging by, 168
  personnel problems of,
    165-168
  restraining of, 126-130
  role in permitting process,
    11-15
Regulatory Procedures Act,
  129
Regulatory reforms
  in the 1970s, 109-142

in the 1980s, 149-196
continuing challenges of,
  211-214
cost-benefit approach,
  127-130
joint review process for,
  130-140
judicial versus
  administrative, 127-130
need for, 103
politicization of, 210
project planning by
  industry in, 168-184
Regulatory Reform Act,
  126-130
restraining regulators,
  126-130
role of government in,
  150-168
role of industry in, 168-184
site selection by government
  agencies for, 123-125
streamlining, 109-115
through better
  communication, 184-197
tips for, 205-211
truncation of the process,
  115-123
see also Permitting process;
  Project planning
Resource Conservation and
  Recovery Act (RCRA), 162,
  163
Rhode Island
  environmental controls
  rank, *225, 226, 227*
Rivers and Harbors Act, 39
Rosenthal, Christian, 89-95
Ruckelshaus, William, 169

S

Schedules, project decision,
  159-161
Scoping, 34, 40, 42, 189
Sherman, Harris, 131
Sierra Club, 114

Site selection, by overriding agency, 123-125
SOHIO, 35, 43, 87, 89, 95, 157, 166, 174, 181, 186, 192
   case study of, Pactex project, 26-35
   see also Pactex project
Solano plant. *See* Dow Chemical Company
South Carolina, 5-7, 59, 113, 122
   environmental controls rank, *225, 226, 227*
South Dakota
   environmental controls rank, *225, 226, 227*
Standard Oil Company of Ohio. *See* SOHIO
STEAG
   case study of plant siting by, 95-100
   one-man opposition to, 96-100
Storm King Mountain power plant, 182-183
Strauss, Robert, 52
Surface Mining Control and Reclamation Act, 192
Super-siting. *See* Permitting process, truncation of

T

Tellico Dam project, 119
Temporary Emergency Court of Appeals, 180-181
Tenneco Corporation, 6
Tennessee, 3, 4, 113
   environmental controls rank, *225, 226, 227*
Tennessee Valley Authority, 194
Texas, 3, 8, 34, 59, 165-166
   environmental controls rank, *225, 226, 227*
Thurmond, Strom, 6

Thurow, Lester, 213
Tidewater Refineries Opposition Fund, 36
Toxic Substances Control Act, 162
Tracking systems, project, 161-162
   PEP, 161-162, 208
Trade secrets, 187-189
Trans-Alaska pipeline, 27, 30, 81, 118

U

Udall, Morris, 50
Unilever, 1, 2
Urban Systems Research and Engineering, 85
U.S. Army Corps of Engineers, 20, 25, 36, 38-39, 40-41, 83, 157-158, 191
U.S. Chamber of Commerce, 207
U.S. Department of Commerce, 36-37, 50, 52-53, 68
U.S. Department of Energy, 29, 33
U.S. Department of Housing and Urban Development, 60, 70
U.S. Department of Justice, 188
U.S. Department of the Interior, 116, 117
   see also Bureau of Land Management; Office of Special Projects
U.S. Environmental Protection Agency, 12, 13, 22, 25, 36, 40-41, 49, 57, 62, 83-84, 110-111, 157-158, 161-162, 184, 187, 191, 205-206
   budget cuts in, 166-168
U.S. Fish and Wildlife

Service, 37, 38, 83, 157-158, 164-165, 193
U.S. Forest Service, 130, 135, 152, 175, 193
  rating of CJRP by, 137-140
U.S. Steel, 84, 153-154
Utah, 166, 191
  environmental controls rank, *225, 226, 227*

V

Vail Associates, 130-131
Vermont
  environmental controls rank, *225, 226, 227*
Virginia, 59, 187-188
  environmental controls rank, *225, 226, 227*
  and HRECO Portsmouth refinery, 35-43
Virginia Bureau of Shellfish Sanitation, 38, 41
Virginia Council on the Environment, 41
Virginia Department of Health, 38
Virginia Institute of Marine Sciences, 37, 38, 41
Virginia Marine Resources Commission, 36, 38

W

Washington, 195
  environmental controls rank, *225, 226, 227*
  super-siting agency in, 119-120
Washington Environmental Coordination Procedure Act, 110, 155
Washington State Department of Ecology, 110
Water Pollution Control Act, 50
Water Quality Act (1965), 6
Watt, James, 117
Weidenbaum, Murray, 43
West Virginia, 61
  environmental controls rank, *225, 226, 227*
Western Fuels-Utah, Inc., 179-180, 192
Westinghouse, 180
Williamson Act (California), 20
Wisconsin
  environmental controls rank, *225, 226, 227*
Wyoming, 134, 180, 183, 196
  environmental controls rank, *225, 226, 227*
Wyoming Industrial Council, 191

ABOUT THE AUTHOR

Christopher J. Duerksen, an attorney and Senior Associate with The Conservation Foundation, has been the principal investigator for the Foundation's four-year Industrial Siting Project, which has studied how environmental laws affect new industrial development and the ways that states and corporations are working to improve the environmental regulatory system. He is the author of *Dow vs. California: A Turning Point in the Envirobusiness Struggle* and the editor of *A Handbook on Historic Preservation Law*. Also for the Foundation, Duerksen directed a project identifying key issues in the management of low-level radioactive waste, which resulted in the report "Toward a National Policy for Managing Low-Level Radioactive Waste," and currently serves as co-project director for a study updating the 1973 book, *The Use of Land: A Citizens' Guide to Urban Growth*. He is also a principal contributor to the Foundation's study on the future of the National Park System. Before joining the Foundation, Duerksen worked with the Chicago law firm of Ross, Hardies, O'Keefe, Babcock and Parsons on projects involving land use and environmental law, wetlands, and historic preservation. Prior to that, he was resident staff consultant in England for The Conservation Foundation's International Comparative Land Use Project. Duerksen holds a law degree from the University of Chicago Law School.